BEYOND THE OASIS

SAFARIS OF SONG AND STONE

Jeannette Hanby
David Bygott

Also by Jeannette Hanby and David Bygott

Spirited Oasis
Lions Share
Kangas: 101 Uses
Ngorongoro Conservation Area: Guidebook
Kilimanjaro National Park: Guidebook
Gombe Stream National Park: Guidebook
Bwindi and Mgahinga National Parks: Guidebook

Also by David Bygott

David Bygott's Gnu Book: A Light-hearted Look at the Gnu, Or Wildebeest
David Bygott's Birds of East Africa: An Unreliable Field Guide

BEYOND THE OASIS

SAFARIS OF SONG AND STONE

Kibuyu Press
3005 North Gaia Place,
Tucson, AZ 85745
USA

www.kibuyupress.com

Copyright © 2021 by Jeannette Hanby & David Bygott
First Edition - 2021

Cover design by David Bygott
Illustrations by David Bygott

All rights reserved

No part of this publication may be reproduced in any form, or by any means, electronic or mechanical, including photocopying, recording, or any information browsing, storage, or retrieval system, without permission in writing from Kibuyu Press

ISBN
978-1-7364953-2-2 (Hardcover)
978-1-7364953-1-5 (Paperback)
978-1-7364953-3-9 (Kindle)
978-1-7364953-4-6 (ePUB)

1. BIOGRAPHY & AUTOBIOGRAPHY, TRAVEL

DEDICATION

To all our neighbors in Mangola, who welcomed us into their community, helped us in many ways, and taught us valuable lessons about village life.

TABLE OF CONTENTS

CAST OF CHARACTERS — IX

MAPS — X

PROLOGUE — 1
Monkey Mischief

CHAPTER 1 — 5
Basso

CHAPTER 2 — 13
Visits

CHAPTER 3 — 23
Whiskey with Wazaki

CHAPTER 4 — 35
Typical Safari

CHAPTER 5 — 59
Kibuyu Partners

CHAPTER 6 — 69
Tea with Tomikawa

CHAPTER 7 — 79
Bungeda Bashkai

CHAPTER 8 — 91
Gudo the Guide

CHAPTER 9 — 101
Treasure Hunting

CHAPTER 10 — 113
Singing Stones

CHAPTER 11 125

Kolo Rock Art Safari

CHAPTER 12 141

Kichaa

CHAPTER 13 149

Schoolgirls

CHAPTER 14 159

Footprint Man

CHAPTER 15 167

Wildlife Love Stories

CHAPTER 16 181

A Fool Safari

CHAPTER 17 193

Robbers - El Niño

CHAPTER 18 201

A Day on Lake Eyasi

CHAPTER 19 209

The Millennium Move

EPILOGUE 219

ACKNOWLEDGEMENTS 221

FURTHER READING 222

CAST OF MAIN CHARACTERS

Home crew—**Athumani**, our esteemed cook and advisor in all local matters; **Len**, **Sam** and **Gillie**, our three foster boys; **Pascal**, "Man Friday" who did many different jobs; **Gwaruda**, neighboring pastoralist who was also our watchman; **Ali**, neighboring farmer who became our head gardener. **Issa**, One of Mama Rama's minions who also worked for us, mainly in our garden plot.

Mama Rama—An Iraqw woman central to the village and essential friend.

Basso—A Datoga pastoralist who befriended us after we saved his life.

Matayo and **Abeya**, and their children **Adam** and **Abande**—A Hadzabe family we knew well.

Dr Yoichi Wazaki—Japanese researcher interested in East African cultures.

Mzee Bashki—A Datoga clan leader. His death was marked by a huge ceremony.

Jumoda—Young Datoga man, both friend and enemy, who had a great impact on our lives and tourism.

Dr Morimichi Tomikawa—Distinguished Japanese researcher and medical doctor who studied the Datoga pastoralists.

Gudo—A Hadza friend and companion who guided us to a better understanding of the entire Mangola and Eyasi region and also worked for many different academics and researchers.

Dr. Mary Leakey—Friend and well-known archeologist who discovered fossils at Olduvai and ancient footprints at Laetoli.

H.H., **Kichaa** and other "Hadza Helpers"—A mixed bag of people with ideas about how to help the Hadza live.

Dr. Charles Musiba—Tanzanian anthropologist who studied human footprints.

Dr. Daniela Sieff—British anthropologist collecting data on Datoga pastoralists.

Nani and Chris Schmeling—Friends living on a farm at the edge of Lake Eyasi, developing a campsite and tented lodge by beautiful springs.

Johannes and Lena Kleppe—German owners of Mangola Plantation.

A few names have been changed, to avoid embarrassing the people concerned. Chapter 4 is fiction, though based on years of experience. Hatari Safari and its staff, Fisi Lodge, and the members of David's tour group are all fictitious. However, the places they visit, the animals they see and some of the people they meet are real.

THE FOUR MAJOR MANGOLA ETHNIC GROUPS

These were genetically, historically, socially, linguistically distinct peoples all living in Mangola during the time we lived there.

Hadza—a unique, local and small group of traditional hunter-gatherers speaking a click language.

Iraqw—a lineage of farmers who also kept some livestock, migrating over centuries southwards from their origins in Ethiopia region, speaking an Afro-Asiatic (Cushitic) language.

Datoga—herders of livestock displaced by their relatives the Maasai who also migrated south along the Nile speaking Nilotic languages.

Bantu—a major group with many subgroups, farmers originally from central and western Africa speaking Bantu languages—Swahili being the major one and used by all these groups as well as foreigners.

XI

MAPS

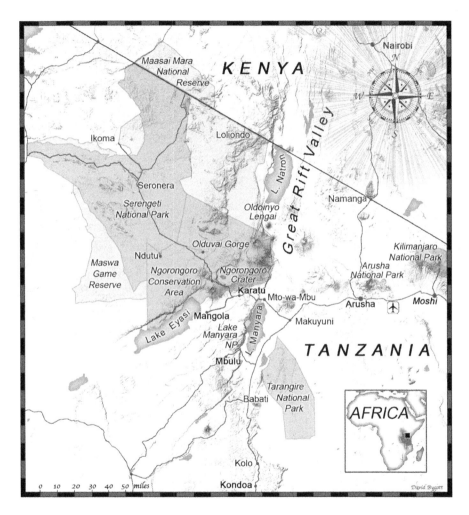

Map of Northern Tanzania. The Mangola area at the east end of Lake Eyasi is surrounded by famous wildlife areas. Serengeti and Ngorongoro lie to the north with Lake Manyara and Tarangire to the east, both on the floor of the Great Rift Valley. Further east, a string of volcanic mountains culminates in mighty Kilimanjaro.

Map of Eyasi-Manyara area. This shows most of the locations described in the stories. Near the center, Gorofani was our "home village" and Mikwajuni was the home that we built, after a sojourn at Flycatcher Farm near Oldeani. Karatu and Mto-wa-Mbu were sources of supplies. Supportive friends lived at Gibb's Farm and Kisimangeda. For wildlife experiences, it was easy to visit Ngorongoro Crater and Lake Manyara National Park.

XIII

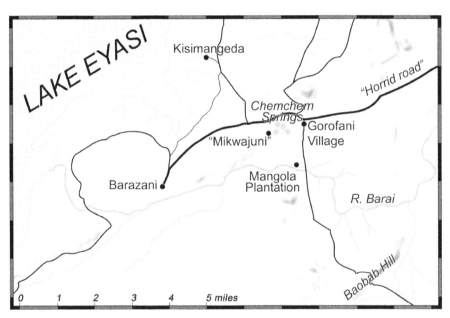

Map of Mangola area. This is the loose name of a cluster of villages including Barazani, Gorofani, and Kisimangeda. Groundwater springs such as those at the Chemchem oasis are the lifeblood of these farming communities. Away from the springs and the Barai flood-plain, the terrain is arid and rocky. Mangola Plantation was the farm where we first lived before building at Mikwajuni alongside the stream.

PROLOGUE
MONKEY MISCHIEF
INTRODUCING MANGOLA

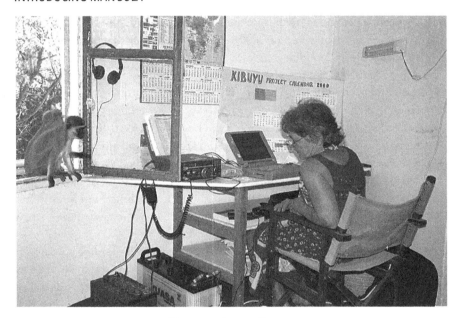

Vervets invade our radio room

Monkey invasion—the vervets arrived with a bang. One dropped from the tree onto our Land Rover parked outside my open window. The metallic clang tore my attention from tuning the radio that let us communicate with the outside world. Welcome to our inside world, a homestead by an oasis in a remote part of Western Tanzania.

More thumps came as silvery-furred monkeys landed on the car. The ruckus almost drowned out the radio noise. The youngsters started wrestling, pushing each other off the top, dangling from tree branches, leaping to the ground. I laughed in delight, so entertained I just about missed hearing the pilot calling on the radio to announce he'd soon land at our village airstrip for the monthly health clinic.

Where is this place with monkeys, an oasis, and flying doctors coming to a village clinic? Mangola is the name. Have a look at the map. Mangola spreads itself around the east end of Lake Eyasi, which normally isn't a lake at all but a dry expanse of soda. The Eyasi Basin is huge, extending from the southern end of the great chain of volcanic mountains known as the Ngorongoro Highlands in the

Great Rift Valley. Lake Eyasi's northwestern steep escarpment separates it from the famous Serengeti.

Why did we live in Mangola? Our first book, *Spirited Oasis*, tells that story. In essence, the never-ending wind from the moist highlands pushed us into the arid lowlands while we looked for a place to live. At first, we hid out on a sunflower and onion farm. After some years there, we left to explore more of Tanzania. We returned to the land of playful monkeys when the village council invited us to choose a plot of land.

We chose ten acres with shady trees near freshwater springs called Chemchem. A papyrus-lined permanent river from the oasis separated our plot from the village. Many years went by while we built our round houses. During that time, we tried to survive financially by producing artwork, books, cards, and displays, plus guiding safaris. We acquired various workers, loose children, and visitors. All of us, including the wildlife, shared the place, the spring waters, and shady trees.

Living in the outback involved hard physical work, establishing friendships, and learning social rules. Our village had no hospital, post office, filling station, or bank but did have a fabulous landscape and fascinating peoples. Living on our own resources wasn't easy for us foreigners. Having the radio that the doctor pilot installed for us helped. Not only did medical news arrive; we used the radio to keep in touch with distant sources for work and support. Reaching our closest town demanded a rocky and dusty or muddy ordeal up the track we named the Horrid Road.

But today I had no car to drive on any road. David was away leading another safari. I aimed to walk over to the village airstrip to greet the pilot and medical crew. Investing time into these friendships was my personal health insurance policy. The monkeys disappeared suddenly as I left the radio room. Did I scare them, or was it the big, bad baboons muscling their way into our compound? They even scared me. I double-checked that I'd closed the window and door.

Heading to the kitchen area, I greeted various staff and visitors who sat, eating or working, under the twisted thorn tree in the center of our compound. Athumani, our cook, butler, and support system, gave me my basket of goodies that I'd prepared. The walk to the airstrip started along the streamside, under the fragrant fig and yellow-barked acacia trees. I crossed the head of the oasis where many little seeps and springs started their journey through the papyrus swamp.

A youth lounged outside a tourist camp and waved. "Habari yako?" I called to Jumoda in Swahili. In English, that means Hello, what's your news? We all used Swahili as our common language, regardless of tribe. Politeness and greetings were mandatory. All along the way to the village, I doled out hellos to everyone. I encountered members of the Datoga tribe with their cattle, some foraging Hadza hunting baboons, many Bantu working on their farms, and a few Iraqw heading to the village with goats. The mix of different tribes was part of what attracted us to Mangola. Finally, I reached the cluster of houses that made up Gorofani village, then the airstrip. "Hello," I said to Pilot Pat and gave him and his nurses tea and cookies.

And so began a day of encounters with wildlife and different peoples from many places and backgrounds. Interactions with people in the Mangola landscape are the basis for this book. In our previous book, *Spirited Oasis*, we introduced individuals you will meet again in this book. But here we want to share with you the adventures that extended beyond our oasis.

Young vervet monkeys play

What of ourselves? David is a born naturalist, nurtured in the fields and forests of England. Jeannette grew up in suburban California among disappearing citrus groves, hating the effects of "progress," always longing to escape to wilder places. David and Jeannette met in Cambridge at the Sub-Department of Animal Behaviour. We share an enduring interest in our primate heritage and the natural world in all its writhing wonders.

Our interest in how animals live in groups led us to study lions in the wilds of Serengeti. Then we moved into conservation work, based in Tanzania cities. A dislike of urban life and stressful events propelled us to settle in a remote part of Tanzania. And so, we ended up in Mangola.

Like the vervet monkeys bouncing off the Land Rover, we bounced around the land, exploring, discovering, and learning. And also like the monkeys, when they saw baboons or smelled a leopard, we sometimes moved silently and carefully.

Read on for stories of fascinating people, adventures to special places, tales of woe and wonder, and safaris in space and time.

Vervet sketched on our front porch at Mikwajuni

CHAPTER 1
BASSO
AN EMERGENCY, A DATOGA WARRIOR, A FRIEND: 1989

"A lion bit him." That is what the worried father said.

The story of the lion bite started on a morning when the local troop of vervet monkeys swung and played outside the window. Watching them tickled me; I couldn't concentrate. I had sums to tally, a radio message to send, and a supply trip to organize. David and I were getting ready for a trip to Karatu town, and that meant preparing ourselves for what we called the Horrid Road. It would be a long, bumpy, dusty, rocky trek from lowland Mangola to the highlands. We had lists to make, boxes and merchandise to load, and the car to check over.

As I closed the windows on the distracting monkeys, my ears caught the sound of agitated human voices. The babble came from the center of our compound—the *baraza*—our meeting place and outdoor kitchen. I sighed deeply. I most definitely did not want to have to deal with strangers on this busy day.

Calm down, I reminded myself. Remember, oh impatient one, you chose to live here in rural Tanzania. No one forced you. Part of living in outback Mangola demanded becoming involved in the social scene. I had a foolish notion that my encounters with neighbors and strangers would help turn me into a more sociable, diplomatic person. Several stories in this book show all too vividly how I failed in the endeavor. Even so, I tried.

I peeked out the door to see what was going on. The soothing tones of Athumani, our cook-cum-butler, stood out from the voices of several men. I couldn't see the people and hoped that Athumani could cope with the situation.

Opening the door, a bit wider, I saw David coming down from our entry gate, probably opening it for our exit. He saw my face and looked over at the kitchen area. He said, "It's your show, Mama Simba, those men at the kitchen. I have to get the Land Rover tires fixed before we go."

"Sure," I said, knowing he was even more reluctant to deal with strangers than I. And yes, I mostly preferred dealing with people than having to change tires or fix car engines. Gathering my kanga cloths around me to hold in my courage, I left the radio room and crossed to our house. Sneaking into our bedroom, I used my binoculars to peer down the path to the center of our compound.

I didn't recognize the three men standing at our outdoor kitchen table. They wore *shukas*—black cloaks adorned with fringes and buttons—marking them as members of the livestock-keeping tribe called the Datoga. They stood tall and turned, their eyes following Athumani as he walked up the path towards me.

Athumani stopped at the door and knocked. I opened it slowly. He looked at my frown and said in Swahili, "Sorry, Mama Simba, it's an emergency. You're needed."

Reluctantly, I went to hear the bad news. A handsome, older man greeted me in Swahili, the language all of us shared.

"Can you help my son?" the elder implored. "He is very sick. He needs to go to the hospital."

I'd heard this request too often to scurry. I needed details.

"What exactly is wrong with your son?" I asked.

"He has been hurt. Two days ago, warriors brought my son home. He is badly wounded," the father told me, twisting his walking stick in his hands.

"Why has it taken you two days to come here?" I asked.

He explained that his *boma*—his homestead and corral—lay beyond Endamagha village, at the bottom slopes of Oldeani Mountain many miles north of us. "We had to find several young men to carry my son on a stretcher. We left before dawn and have been walking all day."

In his firm and tired voice, he added, "My son's name is Basso."

Did the father expect me to know this fellow? Basso's father paused, leaning on his stick. I looked at the father and saw worry and exhaustion on his face. My mind started picturing the possible injuries the boy could have. Snakebite? Knife wound? Fell off a cliff?

"What sort of injury has Basso got?" I asked.

His answer shocked me, "A lion bit him." My imagination leaped just like the predator. As a former lion researcher, I saw the attack play out in my head—Datoga warriors, young men moving through the bush in secret, hunting lions against the law. They creep through the canyons of wild Oldeani Mountain. A lion is spotted, they surround the beast. Spears raised, the warriors hurl them at the lion. The wounded and desperate lion turns to face its enemies, then attacks.

I snapped back to reality—yes, this is a real emergency. What about infection? The injuries would be severe enough, but the bacteria in a lion's bite could be even more dangerous. Excusing myself, I went to grab my precious book, *Where There is No Doctor*. I skimmed the pages. I knew there wouldn't be advice on lion maulings, so I looked up what to do about punctures, bullet wounds, and infections. The book gave me ideas of what I could do to help.

Basso portrait

"We will take Basso with us," I told the father. He stood straighter, and his face relaxed a little.

"We'll go to the hospital at Oldeani village as soon as we can. Where is Basso now?"

"He is near the house of Mama Ramadhani, by the main road at Gorofani. He's lying by the roadside with the other warriors." The dirt and noise of the main road would add to the youth's trauma.

"Please," I said to the three men, "go over to the village and tell Mama Ramadhani to give Basso clean water to drink. Tell the warriors to be ready to put him into our car."

By now, Pascal, our do-everything man, and our three foster boys, Len, Sam, and Gillie, had joined the group gathered around the table. Time for me to delegate tasks. David was getting the car ready, loading the usual equipment we took for every trip—boxes, baskets, spare tires, and so on.

I sent Pascal and the three boys to collect mats, a mattress, sheets and blankets for the wounded Basso. I told Athumani to boil water while I got my medicine box. My little cabinet contained bandages, wound dressings, sulfa powder, antibiotic creams, and pills. I wasn't sure how to treat a young man mauled by a lion, but I would try. Together, Athumani and I gathered buckets, basins, and cloths, and stowed it all in the car.

David drove quickly out of our compound. When we reached Mama Rama's place along the main road, we saw the three Datoga elders just arriving. They joined a group of six black-cloaked warriors clustered like a flock of vultures around a body lying face up on the ground. I hopped out of the car and stared at the tall men hiding the prone figure.

His father pushed the youths aside with his herding stick so I could get closer. Basso looked terrible, his eyes crusted and fluttering, his skin the dusty color of old ashes. His thin body trembled, shook, stiffened, relaxed, then shook again. I listened to him wheezing, like a leather bellow with a hole. I felt his hot forehead. Fever. He probably had an infection.

I asked the men to remove the shuka. Basso moaned as they turned him, revealing his chest with dried and fresh blood oozing from two holes in his side. I felt sick at the sight. Basso's wheezing told me he could have a punctured lung, perhaps broken ribs too, from the impact of the lion or the fall. I pulled aside the corner of the shuka that covered Basso's lower body and saw the appalling bite wound in his groin. Though deep, the lion's teeth hadn't hit an artery because the blood flow was small, mostly dried and caked. But the area was swollen and hot to the touch. If Basso survived, would he be able to father children? I whispered to myself, *Just worry about keeping him alive for now.*

As I cleaned and bandaged the wounds, I considered the wisdom of giving Basso an antibiotic to cope with the infection. I'd not been giving sick people such medicine because the government could jail anyone for dispensing drugs without a license. I was no doctor. If I gave Basso my ampicillin tablets and he died, I could be imprisoned. I decided I'd take the pills with me in case I could give them to a proper doctor who might need them to treat Basso. Meanwhile, we'd try to rehydrate him and get him to the hospital.

I told the father and warriors to wrap Basso in a clean sheet and get him to drink more water. They carried Basso to our car and laid him gently on the mattress we'd put in the back. His father, another elder, and a warrior climbed in and hunched beside the young man. The rest of the grim-faced men stood and watched us leave.

Off we went, up and up the stony, corrugated road, Basso groaning at every bump. I knew that this lion-hunting warrior held in his pain, not giving voice to the agony he must have been feeling. Young Datoga men grow up expecting to endure any pain—injury, illness, even circumcision—with stoicism.

David drove as fast as he dared. After a harrowing drive, we arrived at Oldeani Hospital. This crumbling establishment sat glumly on the slopes of the mountain that gave both the nearby village and hospital its name. We followed a gravel drive to the front of a long, low building. Once upon a British time, the hospital might have been able to cope with such an emergency. But on that day, the place was as unkempt as the patients who wandered about with their families among un-swept leaves, prickly pear cactus, and dead weeds.

We asked several questions to blank or befuddled faces before we found a helpful male nurse. It took even longer to retrieve a doctor from the dilapidated houses behind the hospital. Basso's father stayed in the back of our car, murmuring reassuring words to his son. The lad showed no signs of consciousness except for groans. We talked to the medical people. Their reluctance to help Basso resembled that of a troop of monkeys encountering a wounded lion. Their excuses included lack of medicines, bed space, and equipment, making it clear that they wanted us to take Basso away. Even better, take him immediately to the hospital in Karatu.

I refused to make the longer and riskier ride, genuinely worried that Basso would die on the way. We argued, and at last, the doctor agreed to take the wounded young man. The three Datoga men gently slid Basso out of the car onto a stretcher. Looking at his thin ashen face, I didn't think he'd survive. I told the doctor how I'd cleaned the wounds and asked if he'd start antibiotic treatment. He claimed the hospital had no medicines. I slipped the packets of ampicillin I'd brought into the hands of the father who'd been intently following our Swahili conversation. He nodded and moved forward to hand the pills to the doctor. I felt a surge of gratitude that the doctor accepted without question. He nodded; I nodded. Deal done.

The doctor, Basso, and the father disappeared into the dark interior of the hospital. The other men went to stand in the shade and wait. We drove on to Karatu, wishing we could have done more.

For weeks, we didn't hear much about Basso's prognosis. The villagers I asked told me that he was transferred to another hospital. We reckoned he might be at the regional hospital in Moshi town by Kilimanjaro Mountain. I never did get the full story of his recovery but when Mama Rama told me he'd survived, I felt relieved. Basso's wounds had healed; he'd returned to his father's boma to recover completely.

I wanted to visit Basso but felt uncomfortable. I didn't want his family to feel pressured by my visit. They didn't know me; would they believe I had a genuine concern about the young warrior? I feared they might think I was acting like a tourist—or worse, fishing for a reward for helping him. So, I didn't go.

Some months later I learned more. On *mnada* day when Mangola's outdoor monthly market was in full swing, I returned from my prowl among the crowds. I had my cotton kanga cloths wrapped over my head to shield me from the sun, dust, and stares. My eyes adjusted to the welcome shade of the twisted acacia tree by our central kitchen where I sank onto a bench. I greeted our guest from Australia, who handed me a very welcome cupful of tamarind juice.

She looked over my shoulder and murmured, "Oh, look. A couple of young gals are following you." I turned to see two Datoga youths gliding towards us. One wore the traditional shuka, the other dressed in a baggy shirt and shorts. I looked behind them for a glimpse of the young girls. Was I too sunblind to see them?

Then I got it. My friend had mistaken the gender of the slender, supple Datoga male body. Many foreigners looked at Datoga men in their flowing, fringed garb that cloaked them neck to toe and saw females. In this case, the two youths did look pretty, with their smooth skin, long eyelashes, and graceful posture.

Baggy Shirt Boy towed a goat. I thought the pair wanted to sell the goat and I'd have to say no. Our three foster boys had gone to the market early and bought several goat legs; we didn't need more. I stood up, prepared to say, "No, thanks." Then I stared. The older of the two smiled, greeted me, and handed over the goat. "Mama Simba, this goat is for you, a gift. It is to thank you for keeping me alive."

Surprised and delighted, I realized this young man was Basso, the survivor! In Africa, if you save someone's life, you become responsible for them. And so, Basso became another of our extended family. When he, too, started a family of his own, I knew he'd totally recovered.

My only regret is that I never asked what happened to the lion.

CHAPTER 1: BASSO

Basso brings a goat

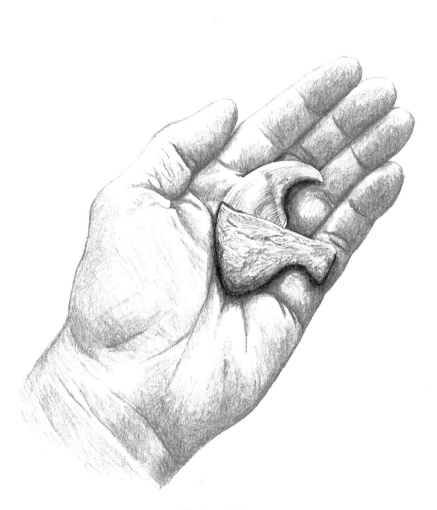

The claw of a lion

CHAPTER 2
VISITS
A SAFARI TO A HADZA CAMP AND A DATOGA DANCE, 1991

Datoga lion-hunting warrior

"Mama Simba," came a call from the kitchen area.

I recognized Basso's voice. He and I were what I call "trauma friends." We shared a bond because of the emergency that nearly claimed his life. Several months had passed since the wound from a lion's mauling had caused us to rush him to the hospital. He seemed to have recovered completely.

Basso and two other Datoga warriors stood waiting to greet me at our communal kitchen table. The soft early morning light coaxed handsome features from their dark faces and black-cloaked bodies.

"This is my uncle," Basso said, nodding at the tall, older man. Basso gave his uncle's name, so unpronounceable that it promptly fled my brain. He then introduced the shorter, stockier fellow. I didn't catch his name either, because I found myself staring at him in awe. A Datoga hero stood before me. I could tell by all his lion killer paraphernalia—a lion claw attached to a thong on his arm, a spear, a decorated shield, many brass bracelets, and rings. Basso smiled at me and glanced at the hero. "Piga picha?" *Take a picture?*

Back then, all cameras used film. Picture-taking meant fiddling with camera settings, maybe adding a flash, carefully storing the exposed film, getting it processed in town, then back safely home. David was our designated photographer, but he was off on a safari with tourists. Ken, our resident Datoga researcher who loved taking photos, wasn't in sight. Basso was my friend; the task, mine. I plodded off to get my camera.

The three men presented themselves with stiff poses and straight faces. Photos of them wouldn't show their liveliness, power, and personalities. I had an idea. Smiling to myself, I went to fish out my old Polaroid camera. With a whirr, the photos emerged, and I delighted in watching the men's amazement at pictures magically developing in front of their eyes. As they amused themselves with the Polaroids, I snapped a few pictures with my regular camera, all three of them relaxed and smiling, talking, gesturing.

Now came my turn to ask a favor. I asked Lion Killer to demonstrate how he'd killed the lion. He stood tall then dropped to a crouch, proceeding to act out a hilarious creeping through the brush, swiveling his head this way and that, keeping track of his comrades as they surrounded the poor lion. Then the attack! Lion Killer held his shield close and thrust with his spear. Standing tall, he stood over the dying beast. He turned to face his small audience with a stoic face. We stared with full attention, obviously impressed with his reenactment of the hunt.

Lion Killer went on to show how he advertised his great feat when he went roaming through Datoga country. When he approached a cattle compound, he gathered his gear and strode forth singing. He ululated into his shield, waving it to and from his mouth to produce an odd wailing song. That haunting sound would have terrified me and maybe even a lion on a starless night in the bush.

Basso's uncle told me in Swahili, "When a man has killed a lion, he goes from homestead to homestead and collects tribute. Each family usually gives him a cow because of his courage in protecting our livestock."

Lifting my open hands to Lion Killer I replied in the same language, "No cows, just the pictures." Even though I respected this fearless warrior, as a former lion researcher, I was on the side of the luckless lion. As a gesture of praise, I give them soap and matches from my supply of little presents. They stuffed the things inside their cloaks. Then they just stood there, silently. I expected them to leave. They didn't.

I looked at Basso, his face blank. Then my mind-your-manners memory kicked in. I hadn't offered them food yet. All visitors are offered food and drink as a matter of course, whatever time of day; it is a first principle of African hospitality.

"Please come and sit at the table," I said, "I will find you something to eat."

Uncle gave me an explanation for their ready acceptance of my offer. "We are hungry because we did not get back to our boma last night. We slept up there." He gestured vaguely to the hills behind us. I put honey on thick slices of bread while Athumani made tea.

While they ate and drank, I walked to the researchers' camp to find our current resident anthropologist, Ken. Since he was studying the Datoga, I reckoned it was his job to look after our visitors. I strode into his camp and found him asleep. "Good morning," I said emphatically in my cheeriest voice. "There are three handsome members of your study tribe at the baraza. If you come right away, you can get some good photos of a lion-killing hero."

Turning away from the groan of a response, I got back to work. I thought my visitor problem solved. However, as the day warmed to hot, I went down to the river for a quick dip and found all three Datoga fellows under a shade tree lying on the grass, covered in their black shukas.

"What happened to Ken?" I asked.

Basso got up and stretched. "He went back to his camp. We are waiting for you."

"Why?" I asked.

"We would like to know when you are coming to see us."

"Am I coming to see you?"

"Yes, Ken told us you wanted to visit us at our place. You are very welcome, Mama Simba. When do you want to come?"

What had Ken gotten me into? Basso and his clan lived along the gullied road at the north end of Lake Eyasi. I was pleased to be invited, but it would be a long day and a body-bashing, bumpy drive. We didn't often visit anyone or anywhere just for the fun of it. Should I go? What would I do there? My mind flapped and fluttered like a bird investigating bits and pieces laid out over this path of possibility.

The Datoga men stood, readjusted their cloaks like ravens' wings flapping. They waited as my mind pecked at the debris there. A morsel revealed itself, a reason for a visit to Basso's compound—a visit by two friends, coming to stay in the very near future. I knew they'd be delighted to go to a genuine Datoga boma. Also, I'd have the satisfaction of a little revenge on Mr. Anthropologist. He'd go along and prove his usefulness as a driver, translator, and barefoot doctor. Yes!

Right on cue, I saw Ken heading to the kitchen. I called out, and we all convened. We agreed on a day. With a smile, Basso said, "That day will be a celebration day at a compound near my parents' place. You all will be welcome." The celebration had a name, *bung'eda*. This was a memorial for an honored elder. Such events had singing and dancing, a festive atmosphere. Plans made, the three men finally left, striding out on slender springy legs as only warrior pastoralists can. As they left, I started planning our safari.

As always, on any trip, I added other stops to the agenda. In addition to visiting the Datoga, we'd search out one of my favorite Hadza families. They lived at a traditional Hadza camp, a cluster of temporary grass huts on the lower slopes of Oldeani Mountain. Pandisha and his wife, Abeya lived there. Abeya was my buddy, the maker of many of the mats and baskets we used. Her son Adam and daughter

Abande would be there too. I'd finally have the chance to take them the bag of empty lidded tin cans I'd saved. All in the camp could use the tins for storing honey, rice, maize, or matches.

Another useful task for the trip would be to get some potatoes from the farmers at the far end of the lake. And thus, I could anticipate a smorgasbord of potential delights. The safari would be a day to see the easy-going Hadza, with gifts to give them, bags of potatoes to take home, lunch in a palm forest, and Datoga dances and festivities.

I spent the rest of the day preparing for my visitors. We knew Nancy and Lucie from our safaris. Nancy was keen on archeology and Africa. Lucie was a businesswoman, the wife of a diplomat based in the neighboring country of Burundi. Both were fun to be with, resilient, and personable. David planned to join them at Gibb's Farm lodge and bring them down to Mangola.

The next day they arrived, ushered in by a playful dust devil. We shared news while unpacking the unexpected pile of goodies that they'd brought. During our midday meal, I told them about the special safari I'd arranged. My less than attractive spiel went like this— "There will be glaring sun, wind, and dirt in abundance. It will be baking hot. And the road is very rough; you will be bounced around. We don't know what we'll find at the Hadza camp or the Datoga event, but we'll go prepared for anything."

. I reckoned that if I prepared them for discomfort, they might be pleased by a more comfortable reality. I didn't need to worry, though; both women brightened at the idea of exploring the area and meeting local people.

The day of our adventure dawned calm as a Buddha's smile. But soon that turned to a frown when the wind woke the dust devils, covering our compound with grit. David appreciated being allowed to stay behind. We piled into the car, Ken driving the old research Land Rover slowly over the countless bumps and ditches, the sandy washes and rocky bits. Well shaken, we arrived at the small government boarding school in the village of Endamagha.

A few teachers and several children— Hadza, Datoga, and Iraqw—stood around outside the school. I spotted my young friend Abande in the flock. With due courtesy, I asked the head teacher's permission to borrow her to show us the way to the Hadza encampment. She seemed eager to come along, but the teacher wasn't so keen to let her go. He finally agreed when we promised to drop her back on our return.

Adam Matayo

Beyond the school and small ranger station, we bumbled over rough tracks until we found the conical grass huts of the Hadza camp. They came out to greet us, looking healthy and happy. The loud voice of Adam boomed from a hut, "Mama Simba, shayamo, karibu." *Hello Mama Simba, greetings, welcome.*

Adam was in good form, laughing and posturing, pestering me. "Have you brought me a new shirt! No! Money? NO? Well, what did you bring me?" He tried to look grim and desolate. I shook my head and laughed. Adam put his hands on his knees and lowered his head, pretending he was going to cry, then looked up at me and guffawed. I loved his playacting.

He and a mob of children ushered Nancy and Lucie around the camp, showing them arrows and digging sticks. Adam's mother, Abeya, took me by the hand and pulled me into her hut. She showed me the mat she was weaving for me. I gave her the orange-red and grass-green pair of kangas I'd brought for her. The hut smelled strange, sharp with that coppery tang of blood. When my eyes had adjusted to the gloom, I noted a large chunk of fresh meat hanging from the beams.

"What kind of meat?" I asked.

"Eland."

Hadzabe people drying meat in hut. Adam is at right.

The eland is a prized meat animal, rare in our area but still living on the bushy mountain slopes. The Hadza camp crouched within the boundary of the Ngorongoro Conservation Area (NCA), where hunting was officially banned. Luckily, NCA officials usually ignored Hadza. The Hadza themselves took care to keep their game meat for themselves. Game rangers and locals might report their hunting activity. Rumors abounded that if caught, they might be sent to jail. They might even die there. Abeya did not need to warn me to keep my mouth shut about the eland meat.

Emerging from Abeya's hut, we joined the rest of the people, joking and teasing in a mix of gestures and Swahili words. Adam again began his act of exaggerated pleas for goodies. "Mama Simba!" he moaned, his hands on his full belly, "I am starving to death! Did you bring me bread?" He looked at the bright sky. "Mama

Simba, such fierce sun! Where is my hat? My sunglasses? And my poor mother, she is making her hands raw, weaving your mats. I will have to carry them to your place at Mikwajuni. I need some money for all that work. Just a little. I beg you, Mama Simba!"

We laughed, especially Adam, and the children giggled. I handed out gifts—a big container of maize, a bag of rice, a tin of sugar, little hotel soaps, boxes of matches, and sewing needles—enough for everyone. The Hadza gave us fruit and water while we sat in the shade alongside their huts. As we said our goodbyes, several people suddenly decided they wanted to go with us. I narrowed the group down to Pandisha and Adam. They could help get the bags of potatoes. I knew their help meant they expected a bag for themselves; fine with me. Everyone squeezed into the car, and off we went, dropping Abande back at school.

We drove through palms thrashing in the midday wind towards the alkaline lakeshore. The Potato Lady came to the edge of her farm, a thin Iraqw woman with the sculpted face of that Cushitic tribe. She led Pandisha and Adam off with my sacks while the rest of us looked for a place to picnic. We settled under a massive, yellow-barked acacia. Undermined by the boggy ground it had fallen partway over. Water from Oldeani Mountain percolated underground until it leaked into the damp edges of the normally dry lake. The groundwater was why people grew potatoes here, building raised beds above the water, no irrigation necessary.

The inevitable gaggle of little boys watched us as we ate our sandwiches and drank my homemade *mkwajuice* (strained pulp of tamarind pods mixed with honey and water). We ignored the boys, but we couldn't ignore a flock of bedecked Datoga. A parade of tall women with regal bearing strode towards us wearing their best fringed leather skirts, capes, and brightly colored kangas. Their regalia included many brass bangles and beaded necklaces. They slowed to exchange greetings. Having an inkling they were on their way to the celebration, I asked them, "Mnaenda wapi?" *Where are you going?* The women confirmed that they aimed to join others at the boma of the deceased elder. An elegant lady in the lead pointed to a distant green palm grove, saying, "Karibuni." *Welcome, come along.* With delight, we agreed we would do just that.

Pandisha and Adam arrived with the freshly dug potatoes, bags slung over their shoulders. They heaved the sacks into the back of the car. We climbed in. Ken started the Land Rover and immediately stranded it on a tussock of rough grass. Gritting my teeth, I suggested that, since he could speak some Datoga, he should escort Nancy and Lucie to the celebration on foot. They set out, following the Datoga women sashaying away into the heat shimmer.

I got Pandisha and Adam to help push, put the Land Rover in low range, and revved the engine. With some to-and-froing, it shot off the mound. We drove cross-country to a dirt track where five Datoga men waved me down. They asked for a ride to the celebration, hardly a ten-minute walk away. Taking a posse of pastoralists to a party seemed the friendly thing to do.

The five men piled in along with an aura of wood smoke and cow dung, jostling and pushing, trying to get their spears and clubs positioned. They would be able to make a grand entrance with all their weaponry if they could get it out without hurting each other, the car, or me.

Pandisha and Adam didn't look at all pleased at this invasion. They scrunched themselves into the far back of the Land Rover among the potato sacks. The Datoga used to consider the Hadza almost on the same level as lions, to be hunted and killed to prove the warriors' bravery. In retaliation, the Hadza had sometimes killed the Datoga with their poison arrows. The enmity between the tribes was still palpable under the surface.

I backtracked through the palm groves, crossing slowly through the bush, when one of the men yelled something. He opened the door and jumped out. I slammed my foot on the brake, alarmed. The car stalled. I looked around at the rest of the men. They shook with laughter.

"Ni mgonjwa?" I asked. *Is he sick*? They shook their heads, laughing harder. I wondered if he'd rushed off to pee but couldn't ask. I started up the car, but one of the men said, "Subiri kidogo." *Wait for a little*. The missing man soon returned dressed in his best black cloak decked out with buttons and metal pieces hanging off the fringed edges. Now I understood. He'd rushed into the bush to get his party clothes.

We rolled on, everyone chuckling and poking at the dressed-up companion. We passed another Hadza encampment. Abande was there, probably visiting friends after school, so I picked her up to squeeze in between a robust warrior and me. I needed feminine reinforcement. The car was now chock full of very high-smelling merrymakers. Their good spirits helped lift me and the vehicle over ditches and through the brush. Finally, the boma came in sight. I parked outside the fenced compound.

We heard the sound of many voices and saw dust rising from the jumping dancers. The Datoga warriors piled out of the car, grabbed their weapons, and joined in the crowd, hooting and calling. Abande and I got out and looked around. Nancy and Lucie sat on stools near a hut on the edge of the compound. Ken stood talking to an older Datoga man. I opened the back of the Land Rover. Pandisha and Adam got out cautiously, looking like chickens at a fox party.

The host was a dapper Datoga elder with a red and white knitted cap. He explained that the celebration was for his father, the departed chief, whose initial funeral had taken place over a year earlier. He introduced me to some of his wives, or sisters, cousins, aunts, or daughters; it was never easy to tell.

I'd brought some gifts—packets of sugar and tea, pieces of red cloth and big reusable water jugs. I gave them to the oldest looking women, and they beamed. Our host offered us some strong honey beer, which we refused. An older woman came and gestured to us to follow her to one of the thatched huts. Lucie, Nancy, Abande, and I joined the women inside. I reckoned the house belonged to the first wife

Mzungu woman meets Datoga women

whom I identified by her age, posture, beads, and dress. I noticed bundles of fresh greenery fastened at the entry, a typical Datoga commemorative motif. Inside, it was shady and cool, free of flies. Women mixed tobacco for snuff and peeled sweet potatoes. They offered us potatoes and sweet tea, which we accepted gratefully.

The women wore their best finery: heavily fringed and beaded leather skirts. Coils of brass adorned their legs. Twists and crisscrossed strands of multi-colored beads hung around their necks. While we admired them, the women scrutinized us, too, in our motley array of shorts, shirts, vests, kangas, and hats. The ostrich feathers in Nancy's hat fascinated them. Our mutual interest seemed a silent acknowledgment of our shared femininity and very different fashion sense.

We started to sample snuff that I knew would make me sneeze, when we heard shouts outside. All the women stood up, abandoning their tasks, pulling on their skirts and robes. The members of the household hosting the bung'eda wore leather strips bound around their wrists, a symbol of respect for the deceased. Outside, we pushed to get places for the start of the ceremony. Men came chanting in pairs and three abreast, carrying bunches of green leaves that they placed at the door of the main house. I smiled to see Basso among the crowd. The band of men I'd brought along had joined too, beating out a rhythm with sticks on shields as they sang.

The dancing began, women forming a line opposite the men. Singing and humming, the young men and girls jumped up and down. Nancy wanted to take photos, and I told her we shouldn't. We'd been invited by Basso and his uncle, not by the lord of the manor. I tried some polite negotiating with our host but sensed his reluctance to allow picture taking. Luckily, Basso asked if we would take a picture of him with his friends, giving Nancy the chance to take photos without seeming to intrude. Then I remembered the Polaroid camera and went to get it from the car.

At the car, I saw Ken in the shade with his medical kit, doctoring children's sores. Good man, I thought. Opening the car door, I found Pandisha and Adam inside, stretched on the seats, feigning sleep. They clearly did not want to join the celebration. I took the camera inside the compound and even our host asked for a picture. The thump of dancing, the din, and dust went on. Ken came in and did his best to leap with the warriors, incongruous in his sneakers, floppy t-shirt, and shorts.

Nearing the time to depart, Nancy wanted to give something special in appreciation of their hospitality. She took the enormous white ostrich plume from her hat and presented it to the oldest wife. The woman backed away as though the feather had been dipped in poison, putting up her hands and shaking her head vigorously.

One of the wives who spoke Swahili said, "No, we cannot take. The game ranger will see it and put us in jail."

Nancy was dismayed, but later I explained that their fears might well be justified. The game rangers might treat the Datoga just like they did the Hadza. With the ostrich feather safely back in Nancy's hat, we offered our thanks and goodbyes. Smiles and waves followed us out of the boma. We'd come as strangers and been treated as honored guests.

Datoga men and girls dancing at bung'eda

Now came the task of maneuvering the car out of the pumpkin patch where I'd unwittingly parked. The Land Rover had a turning radius of a military tank, so I had plenty of time to be embarrassed as I churned back and forth. The jouncing woke Pandisha and Adam, who again climbed into the back seats, then Nancy, Lucie, and Abande climbed in, and Ken squeezed in, too. I had to tell Basso that there wasn't any room left for him and the two girls also wanting a lift. They nodded with smiles and waved goodbye, but I felt a twinge of guilt for not providing transport.

The purpling bulk of Oldeani Mountain loomed in the growing dusk as we headed home. I was worn out, ready for the end of this exciting and exhausting day. Working at being sociable wore me out, yes, but encounters and adventures in the bush also confirmed my love for the people and place. We sped along, and Abande burst into song. Pandisha and Adam immediately joined in, relieved to be heading back, out of Datoga territory, with their bag of potatoes. Whenever two or more Hadza are in the car, they usually sing. Their music always makes me glad. I tried singing with them, wishing I knew their songs better.

We dropped the songsters at the closest camp. In our cheery goodbyes, I again forgot to give them the sacks of empty cans. I regretted it to the tune of rattling tins-of-good-intentions, all the way home.

CHAPTER 3
WHISKEY WITH WAZAKI
A JAPANESE ANTHROPOLOGIST'S VIEW OF MANGOLA: 1987-1997

The hot afternoon wind puffed dust through the floorboards of the Land Rover. I wiped my eyes and sneezed. I was dirty and tired from the drive down the Horrid Road, eager to get home. But first, I needed to stop at Mama Rama's place, her small café, guesthouse, and shop at the village crossroads.

I swerved around a goat standing in her courtyard and slowed to a stop. Levering my sweaty self out of the Land Rover, I had to hop to the ground. The British certainly didn't build Land Rovers for women less than five feet tall. Going around to the passenger side, I pulled out the bag of rice I'd bought for Mama Rama. Hugging the bag to my chest, I turned to see a Japanese man standing smiling at me. I stared at him in his blue shorts, trim jacket, and safari hat. He seemed a vision of neatness in the chaos of the yard.

He held out his hand. Luckily Mama Rama came out of her hut and took the bag of rice from me, saying in Swahili, "Mama Simba, this is Professor Yoichi Wazaki." Wazaki took my dirty hand in his clean one and gleefully pumped it. His English words came in time with each stroke, "Mama Simba, Mama Simba, I hear much about you, Mama Simba!"

"Professor Wazaki, I've heard much about you, too." I knew him as one of the stalwart Japanese anthropologists who came to Mangola to research various tribes.

Researchers like Wazaki always aroused my curiosity. People who set out from their home countries to visit remote places to study other cultures were fascinating. I felt a kinship with them even though I wasn't an anthropologist. I didn't want to spend months detailing and disentangling tribal customs, rules, traditions.

"I want visit you, Mama Simba," he said. "I want to see you and husband. I will tell you about old Mangola. Now you live here, is good you know history this place."

He spoke with plenty of inflection, chopping his short sentences like a sushi master slicing his ingredients. "Now I go to Magugu. You know the place?" He didn't wait for an answer but went on to give me details. "Magugu is south Lake Manyara. In bottom Rift Valley."

I nodded; I knew the place. Flavorful and fragrant rice made it locally famous.

Wazaki continued, "I go to see Mbugwe people. Very interesting. They live surrounded by other tribes. Tribes like Iraqw. Warusha. Gorowa. The Mbugwe peoples, they are squeezed. They try to grow rice and other crops. The Mbugwe peoples interest me; how they survive. Of course, I bring you back good Magugu rice with me!"

Meeting Wazaki thrilled me. He would tell us stories about his work with local tribes. Even better, he could give us more history of Mangola. Here was a walking, talking encyclopedia—if I could decipher his quick cadenced English, with its blending of r's with l's. But his English sure beat my attempts at Japanese.

Wazaki turned to look past Mama Rama. He gestured at two young Japanese men standing in the shade of the building nearby. "These are my students; they learn ways of Bantu people here. We see you later, yes." He opened the door of his dark green Land Cruiser. I smiled because he had to haul himself up, being almost as short as I. The students dutifully piled in, and away they went, almost hitting the same silly goat.

As the dust settled, Mama Rama led me into the coolness of her current favorite hut. Over tea, she told me more about the professor. "Wazaki has been coming to Mangola for many years. He is not like the others who come from Japan. He will not sleep in one of my houses." She pursed her lips and lifted her eyebrows as she gazed at the road and her busy compound. "Maybe it's the noise of the trucks. Maybe I have too many people here, the children, the donkeys...." On cue, a donkey near us let out a loud bray. Mama Rama threw her head back and laughed. I joined the laughter, trying to imagine neatly outfitted Wazaki perched on a stool amidst the chaos and dirt.

"Wazaki takes his tent for camping, up the hill there," she waved at the slope of the hill behind her place and added, "He has more tents for his students." Enigmatically she said, "And in the village is his other house, his woman, and son, Fadhili."

CHAPTER 3: WHISKY WITH WAZAKI

That bit of information surprised me. We knew Fadhili only slightly. Mama Rama explained, "Wazaki built a house for his woman, Fadhili's mother. It's the one near the shop right near the road into Gorofani village. He gave her all sorts of things. She wastes the money."

When we got to know Wazaki better, it puzzled me that he never introduced his son to us. He didn't talk about him or his mother either. We only met her in passing when we visited his house in the village once. Wazaki was a whirlwind of a man and seemed to take his local family for granted, nothing to show off, nor be concerned about. That attitude contrasted with the attention paid to his research students.

He did come back from his trip, students in tow. They arrived in a cloud of dust. Wazaki climbed out of the car and claimed some packages from the interior. Smiling, he strode up to the table where David and I sat eating lunch with our foster son Gillie. Wazaki put a little sack of rice in front of me and bowed. "Hamjambo Mama na Baba Simba!" he said with enthusiasm. "Hatujambo," we chorused. *We are fine.* We added the customary, "Karibuni," the greeting of welcome to a group because we didn't know if the students and driver were to join us, too.

Happily, no. Wazaki waved to the driver with a motion of sweeping the car away. They immediately drove off. According to our "spy" Gillie, who followed the driver and students, they went to Chemchem Springs to get water and wash clothes. Meanwhile, we invited Wazaki to sit with us. He nodded, smiled, and sat, putting a brown bag on the table. Out of it, he pulled some nearly raw, smoky-smelling meat, placing it on the greasy paper bag "Here is good goat meat, please eat," he said in English, pushing the bag at us.

I looked at the hunk of red flesh and declined. David took a bit and nibbled at it. Wazaki then pulled out what I learned was his standard drink, a bottle of whiskey. He placed a pack of cigarettes in front of him and looked at us both with a broad smile, poked a cig in his mouth, and lit up.

I went to get a bowl and cutlery and watched Wazaki out of the corner of my eye. He looked comfortable, chewing away on his goat meat and downing shots of whiskey while conversing with David. His lit cigarette fumed in an ugly ashtray. David had made it from a tin can, folded over the edges and then painted bright red, like a mouth with lips. Wazaki seemed not to notice this attempt to deter smokers at our table. He laid down the meat, took a swig, then a puff. He put a small amount of rice and beans in the bowl and said, "I promised to tell you about old Mangola."

He launched into his tale. I tried to take notes as best I could, but he spoke in a mix of English, Japanese, and Swahili, mixing them like playing a game with both ping pong and tennis balls.

His rendition of local history went something like this— "I came on a motorbike. It was 1963. I was new but a part of the Japanese team. We were all anthropologists." He paused and spoke again for emphasis, "Social anthropologists." I took that to mean he wanted us to understand that the Japanese team came to study real people and their societies. We nodded.

He continued, "Tomikawa and Tomita came first. That was 1961. They came to study the Mang'ati—we call them Datoga now—and their cattle. Tomikawa built a proper Datoga boma—big corral with huts at Chemchem. He studied how the Datoga live, how many clans, how they grow up, marry, brand their cattle. I stayed mostly in a small village," he paused and looked to the southeast, "over there across the Barai River, in Bantu country. Bantu people are my main interest."

Wazaki slowed down enough to put food in his mouth and swallow before plunging on. "When I came first, the road from Oldeani was terrible." He scanned our faces with a coy smile, nodded, and continued. "Yes, *even worse* than now. One time I came on my motorbike. I got a puncture. I could not fix it in the dark. A lion roared! Such a frightful sound! I try to sleep between a big tree and my motorbike. Then came dawn. I heard noise, rumbles, maybe a car approaching? I rushed out to ask for help. *Oh no*! A rhino crashed through the bush right by me!"

Wazaki put up his hands as though resisting the charge of the rhino, his eyes wide with feigned terror. "What happened next?" I prompted.

"Finally, someone came. That guy had tools and could fix tire of motorbike." Wazaki ended this sentence with a frown. His tone changed as he began a lament we heard again and again. "Mangola has changed much now. No rhinos and roads still terrible. Too many people, too much politics, too many trucks, too much cattle."

His frown disappeared as he added, "Still some trees at my camp, though." His eyes roamed our compound, his head nodding up and down. "Good, you protect trees here. But in Mangola now, too many people. They cut trees for firewood, cut bushes, break limbs. Goats, cows, eat everything. Too many farms, too much onion growing." He waved his arm in a great circle delineating the entire Mangola region, raised his bottle, and stabbed at the sky, crying out, "*Shauri ya watu*!" Then he repeated the phrase in English, "The problem is people!"

Wazaki paused again to nibble at the meat and sip the lubricant to get it down. He swallowed and concluded his tale. "Tomikawa, Tomita, myself. We all come to study peoples here. We come back many times."

Looking wistful, he paused, sat up straight, and announced, "Now I must go. I come back later. I tell you more about research then." He must have heard his Land Cruiser growling along our road because it soon appeared. Wazaki left with his whiskey bottle and the remaining meat stuffed into his vest pocket.

He did return another afternoon, driving himself. We invited him to sit on the covered porch at our private house. He declined our offer of beer, pulling out his whiskey bottle. We put an ashtray by his elbow as he lit his first cigarette and settled to listen to his story about why he'd come to Mangola.

"Yes, I came to study Bantu tribes—Izanzu, Nyamwezi, Mbugwe, Chagga, Sukuma. Each tribe has its language, its own culture. Here in Mangola, these tribes are mixed up. They live together and marry each other. They have come to think of themselves as Waswahili people. Instead of a tribal name, these Bantu people share group identity."

CHAPTER 3: WHISKY WITH WAZAKI

I asked Wazaki if people from different Bantu tribes use the term "Waswahili"—meaning Swahili speakers—for themselves? "No," he told me. "But they think of themselves that way. Most have never been to the Swahili coast, so I ask, why do they think they are Swahili?" He paused to for a puff, then a pull from the bottle.

"The Bantu peoples came to group idea in other places, too. At Mto wa Mbu, not far from here, at the bottom of the rift wall, the Bantu tribes also mix. They, too, have a group identity. These people are not Waswahili just because they speak Swahili. The other tribes here—Datoga, Iraqw, Hadza, you, me—we speak the Swahili language, but we don't think we are Waswahili. When people speak different languages yet have a group identity, this is interesting."

David and I began to understand Wazaki's interest in why the Bantu tribes saw themselves as part of a larger grouping. It occurred to me that Europeans spoke many different languages and identified with their nation state as German, British, French, and so on. When they came to the United States, they began to think of themselves as Whites, a group identity. Yes, that was interesting.

Wazaki continued outlining his ideas, reminding us of the tribal diversity in Tanzania. "You know," he said, nodding, "there are more than 125 tribes in this country. All are different in language and customs. The first people here in Mangola lived like Hadza tribe. They hunted, dug for tubers, gathered wild food."

"Next came the Iraqw people. They migrated slowly down the Rift Valley. They finally settled over there in the highlands." He waved his hand south in the direction of the Mbulu plateau. "The Iraqw people brought crops like millet, and goats."

"Next came the Datoga people with cattle. They lived in the highlands until they got pushed here by their relatives, the stronger Maasai."

This time Wazaki waved his hand at Oldeani and the Ngorongoro Highlands looming to the north.

He continued, "Mang'ati, Barabaig, Datoga—all are names for the early pastoral peoples here. They and Maasai are alike. They are Nilotic peoples with cattle, but they do not have a group identity. The Datoga started to mix with the Iraqw people. They intermarry. But each kept their tribal traditions, no group identity."

Wazaki paused then looked at our cook Athumani, a Bantu from near the coast of Tanzania. He smiled and nodded, saying, "Bantu peoples have come more recently to Mangola. The original Bantu peoples came from central Africa." I nodded, thinking of the spread of Bantus to all parts of Africa.

"The Bantu are farmers. Their way of life has been very successful. They became many different tribes as they moved across Africa. But these Bantu have a group identity. Why do they think of themselves as Waswahili, not Tanzanians, or people of Mangola?"

Wazaki ended with, "This problem of group identity interests me very much."

It interested me, too, the feeling of kinship shared between peoples. Shared history and language pulled people together. I figured similar interests and lifestyles might be enough to give people a group identity, such as religion, or educational background.

Some months passed before Wazaki arrived at our compound again. He came with an older Iraqw man who drove the car and a shy young Japanese girl. I invited the three to sit at our outdoor kitchen table while I searched for offerings of food. I found bread, peanut butter, honey, avocados, and radishes from our garden plot.

Wazaki looked at the spread and raised an eyebrow. I guessed that he wondered when drinks would be offered. I laughed, bowed, and said, "Alas, Wazaki-san, all I can offer you is warm beer." I expected him to refuse, but he brightened, nodded, and said, "Yes!" He drank the whole bottle down quickly; I fetched another.

"Washes out road dust," he said with a straight face, then smiled as he added, "but not so good as whiskey."

He leaned back in his chair and looked at his companions, speaking to the girl in Japanese, then the Iraqw driver in that harsh-sounding language. When he turned his attention back to us, I complimented him on his language skills. "Oh yes," he said in English, "I am a linguist. That is my real work." He beamed as he told us he'd written a Kiswahili/Japanese dictionary. "Oh, yes, I tell you. It was much work to do. It took many years. I don't have it with me now, but I'll bring it. Yes, I will bring it when I come back."

And he did. He brought us his book, unwrapping the famed dictionary from its layers of paper protection. The dictionary was huge and illustrated with his elegant line drawings. Published in 1978, it was still in use.

Some months later, Wazaki arrived at our compound, not for a meal or to ask for help fixing his car. "I invite you to my place," he said bowing. "Come for dinner, tonight, at my camp. I will arrive here to pick you up!"

How could we refuse? And what a ride! Wazaki drove his Land Cruiser like a sumo wrestler fighting an opponent. When he had to slow down for a bump or change down to second gear, he grimaced as though it was an admission of defeat. We reached Mama Rama's compound, chickens, ducks, and toddlers scattering as we juddered to a stop. She came out to greet us.

"Mbuzi iko wapi?" Wazaki asked. *Where's the goat?*

Mama Rama replied, "At your camp. I've sent Maria with tomatoes and bananas, too."

Wazaki revved the engine. We rumbled and crashed up the hill to his camp, where he thrust the vehicle into a bush and jumped out. We staggered from the car to survey Wazaki's camp. The twisted, ancient acacia tree still stood like a contorted overseer. Under it crouched a large green tent, resembling a squat hat with the brim turned up, green storage trunks arrayed in ranks alongside. Several empty chairs stood ready around a merry fire.

Maria from Mama Rama's busily tended to pots on the fire. A man stood in the shadows, dismembering the goat. As we walked towards the chairs, we saw the girl students busy chopping onions and tomatoes while the boys sat around a table poring over paperwork. I chuckled ruefully at the typical gender roles.

CHAPTER 3: WHISKY WITH WAZAKI

Wazaki's camp, Mangola

David and I looked at each other, pleased and smiling. The stage was set for Professor Wazaki's vision of life in the bush. And here we were, as audience and participants. We sat by the fire with our drinks—bottles of warm Safari beer. Wazaki sat too, saying enthusiastically, "Kambi safi sana! Hewa safi! Na kimya!" *A very fine camp. Air clean; and quiet!*

I agreed with all that, except for the quiet. I heard the sound of vegetable chopping, the tent flapping in the incessant Mangola wind, and trucks rumbling on the road below. Wazaki drank his whiskey; we drank our beers. Suddenly he stood up and chased the young men away from the table. The goat arrived. Well, not yet the goat itself, just its liver, raw. Wazaki cut it into small chunks in a bowl and added shoyu from a gallon bottle.

"Japanese delicacy—sashimi!" he explained. He added a squirt of something from a tube. "Wasabi, the mustard of Japan. *Powerful!*" I already knew wasabi as an intense, mouth-inflaming horseradish sauce. I didn't like it at all. Wazaki seemed surprised when David accepted some.

The girls offered us a choice of forks or lacquered chopsticks. Without hesitation, I chose the beautiful chopsticks. David, always inventive, had already carved a pair from two small sticks. They were rough and crooked but far more effective than the shiny ones I had chosen. He didn't drop a single piece of the meat and commented that it tasted good—wasabi and all.

"Can you find such a peaceful place in Japan?" I asked Wazaki.

"No. In Japan, everywhere you hear the sound of cars and machines. Here also are many sounds, but they are sounds of nature—wind in the trees, cries of birds, hyenas calling."

We all quieted, straining to hear the sounds of nature, but only heard people shouting and the onion trucks in the distance, rolling past Mama Rama's on the main road.

"Do many people go camping in Japan?" David asked.

"Some do, but camping is *maridadi sana*—very fancy. People have big tents and rugs, even showers. Not like this! Some of these students here," he waved his hand, "they have never camped before."

One girl put her chopsticks down carefully, then asked David in English, "How you first to Africa?" David provided a summary. "I came first in 1969. I studied wild chimpanzees at Gombe with Jane Goodall." The students nodded, many recognizing the name. "After two years, I returned to Cambridge, England, to write up my Ph.D. research. There I met Mama Simba."

He turned to look at me. I squirmed when all heads swiveled to rivet me in their gaze. David continued, "Mama Simba did her Ph.D. thesis on Japanese monkeys at the Oregon Primate Center. She came to Cambridge to do studies on rhesus monkeys. We married and came back to Tanzania to study lions in the Serengeti. Then we did other things; now we are in Mangola."

The girl nodded, "Ahh...very interesting! Chimpanzees and Japanese monkeys. Professor tells us Mr. Bygottu is an American who comes to Mangola to build hotel!"

We laughed out loud. It wasn't the first time we'd heard the rumors about us building a hotel at Chemchem Springs, about the last thing we'd want to do. Virtually no one in Mangola understood what we did for a living in the wilds of outback Tanzania. Only a few friends seemed to grasp how much we loved the landscape, the people we'd come to know, the amusing and puzzling things that happened every day. Here we had chances to get to know a person like Wazaki, a foreigner who, like us, loved the place.

The moon rose above the trees when the stew was finally ready. We dined with Wazaki and Maria at the table. The students sat on the groundsheet and ate by the light of a small lantern. Panic erupted when a little bush cockroach flew into their lamp. The roach fell onto the groundsheet, then spun and buzzed as it tried to get up. The girls all teetered around with cries of distress, throwing their hands in front of their faces. Eventually, a heroic boy grabbed a can of Doom and proudly sprayed the cockroach to smelly death. The reek of bug spray soon pervaded the camp.

Goat stew always tastes like goat stew, alas, and smothering it with shoyu and Doom didn't make it better. Even so, we continued to eat. Or David did. I slowed and stopped, gagging a little at the smell of the spray. Wazaki grinned at us.

After dinner, we expressed our desire to walk home in the moonlight and escort Maria as far as Mama Rama's. But Maria refused, saying firmly, "I can no walk back. May be lions." Wazaki insisted on taking us in his car. He staggered towards the kitchen area, saying, "You mus' take goat legs wid you. My students go on safari, t'morrow. We cannot eat all meat."

CHAPTER 3: WHISKY WITH WAZAKI

We took a bag and apprehensively climbed into the Land Cruiser. Wazaki had consumed nearly a whole bottle of the potent fermented and distilled cashew fruit liquor known as Konyagi. He was already full of high spirits even before he came to get us.

He started the car, ground the gears, stalled, and restarted the vehicle. He revved the engine, tried to reverse, crunched the gears, and threw the car into low range. We went forward into the bushes. The students all came to watch, hiding their smiles and shouting directions and encouragements. Wazaki had to make several moves to get the car out of its bush before we hurtled downhill towards Mama Rama's place. We clutched at the door handles, ready to jump, but survived the short distance between camp and compound. There we slipped out of the car, shouted our thanks over the roar of the engine, and waved goodbye.

Our good times in Wazaki's company came to an end. One busy day Wazaki drove in while we prepared for a trip. He skipped the customary greetings and small talk and got directly to the point.

"You want to buy my big tent? Tent almost new." A mouse of worry began to gnaw on me. Why would he be selling his big tent? Wazaki was spice to me—tangy ginger, salt, and pepper. He made my life more interesting. I had a particular respect and affection for him, and I didn't want him to take his flavor from our social stew. I searched my mind for something to say. So, as humans often do when we aren't able to express our emotions, we turned to practical issues.

"What is the price of your tent?" I asked.

He told me a ridiculously low price. I tried to convince him to charge more. Wazaki refused and told me he wanted to give us his storage boxes, too. That pleased me because they were sturdy, stackable boxes with lids. But my worry mouse continued to gnaw away. Was he packing up and leaving for good? I watched him scanning around our shady compound, still trying to think what to say.

He looked up into the spreading branches of the tamarind tree and said slowly, as if to the tree, "You have composed a very nice place here. It is good you call it Mikwajuni in honor of tamarind trees. The two of you are good to be in Mangola." I didn't know if he meant Mangola was good for us, or we good for Mangola.

He turned back to me and narrowed his eyes ever so slightly. "Yes, this may be the last trip. I have been here 17 times; maybe I'll not come for the 18th visit to Mangola." I offered Wazaki food, but he refused, saying, "Now I am going back to camp to rest. Please come over to my camp later to see the tent."

We arrived to see the big green tent moving this way and that, blown by the afternoon wind like a giant beast that had lost its skeleton. Wazaki emerged and guided us around the ballooning tent, barely held down with many ropes and pegs. We agreed to buy his tent, turned down his offer of whiskey, and headed home. The next day he brought the storage boxes, half-filled packets of tea, and tins of special Japanese foods. I couldn't bear to let him leave without one more invitation, our last meal together.

"Come for a farewell brunch here," I told him. "You can pack up your stuff, then come for food. All of you. Then you are ready to go without having to clear up your camp."

They came: Wazaki, his driver, three girls, and two young men. The long party table by the river was piled with goodies—pots of regular tea, spiced Zanzibar tea, fresh coffee, juices, syrups, jams, and bowls of fruits. After everyone was seated, I brought the first round of pancakes. I proudly passed the platter around, glancing sideways to see if Wazaki appreciated my efforts. Of course not, I thought, seeing his stoic face. But his eyes twinkled as he asked, "Mama Simba, do you have a beer?"

I laughed, thinking he was joking. I shook my head, no. Without hesitation, Wazaki sent a student off to the car. He returned with a bottle of Safari lager, bowed, and placed it before the esteemed professor. Wazaki proceeded to open and guzzle with satisfaction while he nibbled at pancakes in between drags on his cigarette. I told everyone to fill their plates, but leave room for my brunch trademark, golden heart-shaped waffles.

I placed the first batch in front of Professor Yoichi Wazaki. He looked at them with no great delight. But he looked me right in the face with his mischievous smile as he asked, "Do you have whiskey?" This time I laughed wholeheartedly. I went to our house and brought back half a bottle of Glenfiddich. "Ah, very good, *arigato*," he said.

We ate and chatted away in a mix of languages. Suddenly the students decided to take a photo break. The gang headed to the riverside. A frenzy of picture-taking ensued. They snapped pics of each other in the boat, standing on the dock, and hanging on a rope dangling from the fig tree. The students took turns swinging out over the water, screaming with delight. Wazaki was right there too, grasping the line with hands and bare knees, swinging back and forth, face lit up with booze and pleasure.

They left, shouting, "Thank you, Mama Simba. Thank you, Davidi. We are gone now. Goodbye. *Kwa herini*." Wiry, energetic Wazaki—who drank beer and whiskey with his breakfast, smoked as though cigarettes were an essential food, and howled with delight while swinging on a rope over the river—was in his eighth decade when he left us.

Alas, he did not come back. I grew despondent when I read of his death from cancer. Such a character is a rarity. His departure was a loss to me, us, to Mangola, Japan, and the scientific community, though his dictionary remains extant.

Wacky Wazaki reminds me of those of us who live on the edges of our cultures. Most of us fringe people have shared desires—we want to be individualistic, free of cultural constraints, remaining open to exploring people and places. We shared with Wazaki a deep desire for greater depth in understanding of human society and with that came a willingness to spend time, learn skills, and jump through political, and social hoops. We are a jolly, skeptical lot, maybe a bit eccentric? Yes, we fringe people share a "group identity," a thought that pleases me.

CHAPTER 3: WHISKY WITH WAZAKI

Professor Wazaki at the rope swing

CHAPTER 4
A TYPICAL TOURIST SAFARI
A COMPOSITE OF 15 YEARS OF TOURIST SAFARIS, BY DAVID

Road across the Serengeti plains

"Safari njema!" *Have a good trip*! Athumani called to us, waving goodbye as we drove up the drive, aiming for the Horrid Road and adventures beyond Mangola.

Most people think of a safari as an adventure to see wild animals in wild places. "Safari" simply means a journey or trip, by foot or by vehicle, to town or around the world. Safaris were our life. We made regular supply runs to Karatu town, or less often to Arusha, or rarely to big cities, like Dar es Salaam or Nairobi. We planned these safaris minutely, with elaborate lists, containers for food and fuel, our books and cards to sell in gift shops, and medical and tool kits in case of trouble.

We were freelance safari guides. Tour leading was never a career goal, but our background in wildlife research and teaching suited us to the job. In the 1980s, Tanzania was rebuilding its tourism industry, so had a demand for knowledgeable local guides. I was more tolerant than Jeannette of rough roads and oft-repeated questions. That is why I was often away while she had adventures or disasters in my absence.

This story is about a "normal" safari—not a mass tour nor a luxury trip—set in the early 1990s. It's an imaginary one, taking a real itinerary, fictitious guests, and anecdotes from various safaris and blending them. I describe a group of nine people organized by a small company in the USA and outfitted by a small company in Arusha. Our group had the use of two Toyota Land Cruisers, custom-built to seat seven guests each, with pop-up roof hatches for shade and an all-round view. Our ten-day trip included visiting some of the most beautiful wildlife areas in northern Tanzania.

We were tired and dirty when Mama Simba and I reached the big town of Arusha. That was where most safaris began heading for the Northern Tanzanian Parks, including Tarangire, Manyara, Ngorongoro, Serengeti, and Kilimanjaro. We refreshed ourselves by staying overnight with friends. The plan for the next day involved two different safaris. Jeannette would join me only for the first night with my American guests, to give an "overview" talk. After that, she was eager to join a horse-riding safari with a group of mostly local people. I would be taking a group around what we called the Northern circuit. Here is a rendition of that safari, composed of memories and semblances of our many different safaris.

Day 1. The arrival and orientation of safari members

First, we visited Hatari Safari, our local tour organizer. The name refers to the classic 1962 John Wayne movie, filmed in and around Arusha. We greeted the tour manager, a balding, worried-looking man behind a cluttered desk.

"Welcome, Mama Simba, welcome, Mzee David! Please have a seat. Maria, bring us coffee—oh, and call in the drivers, they're checking their cars."

Maria woke from behind a clunky desktop computer monitor, glared at us, and slouched off into the back. Our two drivers walked in. They looked competent, even in their greasy mechanics' overalls. Adam was a cheerful chatterbox with a silver cross on a chain around his neck. Sakala, tall and dignified, was older and wore a Muslim cap. We'd worked with him before; I was glad he'd be the lead driver on this trip.

We went over the trip details for an hour. Before we left, we agreed on communications. The drivers could talk with each other and with home base by radio when it worked. Mobile phones and the internet were still in the future.

We did chores in town, then drove to Fisi Lodge on the east side of Arusha. Jeannette stayed at the lodge to work on her slide talk while I went to receive my guests at the airport. Sakala met me with a small bus that could hold the whole group. He looked neat and professional in the standard uniform of a khaki shirt, long trousers, and safari vest. I checked that he had drinking water for everyone. The airport involved a 45-minute drive east along a narrow two-lane highway. We passed through lush farms of coffee and bananas, then drier plains of maize and grassland. As the sun lowered behind the massive bulk of Mt. Meru to the west, the snowcap of Mt. Kilimanjaro hovered over the east ahead, a pink cloud fading to purple. A good omen, I thought.

Kilimanjaro International Airport lay on the broad plain between these two mighty mountains. It was small for an international airport but quite pleasant, with 1970s curvaceous concrete and local wood. Its gardens were bright with flowering trees. Sakala and I joined the crowd of drivers and guides with their signboards, waiting for the flight from Amsterdam. The big tour companies, such as Megaglobal and Uppercrustly, were all there, as well as some lesser-known delights like Anti-Pollution Holidays, DoDo World, and Human-like Travel. One company logo boasted, "We go to every measure...to bring you wildest pleasure!" Our board simply said "Hatari Safari," which drew sniggers from the other guides. "Hatari" in Swahili means danger.

At last, the passengers emerged from Immigration into the baggage claim. Through the glass, everyone looked the same. The color was khaki; the jackets had

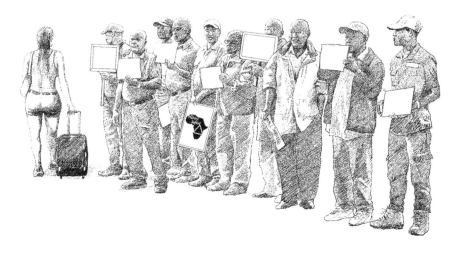

Welcome committee at Kilimanjaro Airport

too many pockets, the faces were weary. They had the right luggage for the rigors of the road, either durable ribbed metal or soft, yielding duffel bags. Some of the travelers likewise would be rigid and impervious to new ideas, dutifully towed abroad by a more enthusiastic partner, while others would bend, flex, and expand to accommodate every experience.

I welcomed six identifiable guests—two singles, Barbara and Peter; two friends, Betty and Jane; and a married couple, Ray and Marilyn. I sent them to the bus with Sakala. Waiting again at the exit lobby, I watched a grim-faced father arguing with two sulky teenagers standing by a pile of luggage. I groaned inwardly, already worried that John, the father, and his kids, Danny and Lisa, would be trouble. I waved them over. "Welcome to Tanzania! I'm your tour leader, David. Everyone else is here. Got all your bags? Amazing!" I ushered them into the bus, and we headed out.

I passed out water bottles. "You won't see much because, hey, this is Darkest Africa! No, you won't see Kilimanjaro because it's now dark, but we may be lucky tomorrow. Try to catch some rest on the way to the lodge."

Sakala turned off the inside lights and people chatted quietly or dozed.

At the lodge, Jeannette met us with keys and instructions. I told them to settle in but get back if they wanted some supper. Otherwise, we'd have our bags packed before breakfast at 6:30 a.m. and a short orientation talk at 7:30.

In the dining hall, the hosts had spread out a nice supper of ham, cheeses, and bread. The guests dribbled in.

"This place is amazing, so much better than I expected!" Peter enthused. Good, you're easy to please, I thought.

"Yes, it's one of the best," I lied fluently. "Just opened last year."

Barbara wasn't so enthusiastic. "I had a spider in my room. Had to call housekeeping to remove it. Will there be bugs tomorrow?"

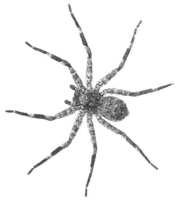

"I hope not. But bring your repellent just in case."

Nobody lasted long, and we were soon all asleep.

Day 2. Arusha to Tarangire

First thing in the morning, the Troublesome Trio had a crisis.

John came straight to me without a greeting. "My wallet has been stolen! I put it on my dresser, right by the bed, before dinner. Then I worried that staff might see it, so I put it under my pillow. But it wasn't there later last night!"

"Dad turned the room inside out, at like midnight!" reported Danny. Lisa, plugged into her Discman (what kids used before iThings), rolled her eyes.

"I'm sure it will show up," I reassured him. "We've never had any trouble here before." This was true, but then, we'd never stayed at Fisi Lodge before, either.

I asked the manager to try again.

After our early breakfast, we gathered in the hotel lounge. I introduced myself, then "...our guest lecturer, Dr. Jeannette Hanby, locally known as Mama Simba."

Jeannette gave a brief slide show, an overview of the Great Rift Valley, and the safari itinerary. I had tricked her by sneaking some extra slides into her show, photos of Mama Simba jacking up our Land Rover next to a lion pride, hauling lion cubs out of an unguarded den, and other daring deeds. She glared at me and hurried through them, but I knew she was proud, as well as embarrassed. The others were impressed.

My guests looked enthused and eager. I asked them to gather their luggage and come to the parking area where Sakala and Adam waited by the cars.

Jeannette, driving our own Land Rover, waved goodbye, telling me, "Safari njema" with a wink and a nod at The Family members arguing in the car park.

Then the lodge manager rushed out, shouting, "We found it, your wallet! It was twisted up inside your mosquito net." John grabbed the wallet, muttered awkward thanks, turned, and got into Adam's car.

The outskirts of Arusha were crowded with cars, buses, handcarts full of cement bags, cyclists, women carrying baskets on their heads.

"You don't see many overweight people," said Barbara, with an expensively trim figure. "What do they do for exercise here? Do they work out?"

"No. They work." I pointed to a woman laboring beneath a load of firewood and wondered why so many people leave their brains at home when they pack for safari. Be kind, David.

Soon we were crossing verdant plains where red-robed Maasai herded their livestock or rode bicycles along the road. The narrow tarmac had no markings except for potholes, which we dodged. The road undulated through bushy hills and small farms, slowly dropping to the grassy floor of the Great Rift Valley.

After two hours and one flat tire, we reached a road junction where a brightly painted metal elephant stood, pointing left. The elephant wore a safari jacket; his red bag said "Tarangire Safari Lodge"—one of my art jobs. As we entered the park, the landscape became fortified with immense baobab trees, their fat, grey trunks stripped of bark up to elephant height, their rounded crowns in full leaf-like giant broccoli. Elephants were there, alone or in groups clustered in the shade, and we stopped to appreciate their ponderous grace. Herds of impala flickered like orange flame, giraffes glided among acacia trees, and black herds of buffalo lounged in the tall grass.

We reached the Tarangire National Park entry gate, a small building, and a metal gate across the road. Danny and Lisa bounded out of the car with their point-and-shoot cameras and headed for a giraffe browsing nearby.

"Oy!" I shouted. "Never get out of the car unless we say it's OK. This giraffe may look tame, but she's a wild animal, and you don't know how she'll react. And there are more dangerous ones out of sight."

Our lodge spread along a ridge overlooking the river. As we pulled in behind the main building, our friend Annette ran out to greet me with a hug. "So good to see you! *Karibuni sana!* You're just in time for lunch. Let's get you checked in. I saved you the tent you like best!" Her husband Big Jon emerged from the workshop, wiping hands on a rag. His blond mustache twitched slightly in hello—the Minnesota version of an enthusiastic hug.

"Lunch is as soon as you're ready," I told my flock. "It's a buffet." Staff in colorful shirts welcomed us, shouldered our bags, and led us along a sandy path shaded by acacias and baobabs. The large tents stood on concrete platforms under thatched

Warthogs grazing

roofs, in a long line facing the river valley. Ahead there was a pile-up. Everyone wanted to photograph four warthogs grazing unafraid between two tents.

"Ugh, those things are gross!" said Lisa.

"Well, I think they're kinda cute," replied Jane.

"Now isn't this Up Close and Personal? Only a mother could love that face!" opined Barbara, the cliché queen.

Lunch was a delicious curry with all kinds of side dishes, a welcome treat for me after Athumani's beans and rice. A waiter patrolled, keeping vervet monkeys at bay while colorful superb starlings sneaked in through gaps in the window screens to steal bread.

After a siesta, we regrouped at the cars and set out for a drive along the river. Immediately, we found elephants drinking and wallowing in pools. We dawdled on to see lions sleeping in the shade, young male giraffes neck-fighting, impalas bounding athletically across the road, and baboons grooming, playing, fighting, and feeding on roadside flowers. The group was well pleased. Happy campers returned to the lodge for hot showers and cold drinks.

In the evening, I took over a corner of the lounge and rigged a sheet for a slide show about predators. Hyenas are versatile scavengers and formidable pack hunters. Cheetahs are speedy but solitary, wimps who often lose their kills to more powerful carnivores. Leopards are "always spotted but seldom seen"—the ultimate stealthy night ninja. Lions are the only truly social cat and our forte, as we spent four years studying them in the field. "The family that preys together..."

During my talk, wildlife sounds drifted up from the valley. "We may hear lion roars in the distance tonight," I told them. "It's not like the MGM lion, snarling when someone pulls its tail. It's more like this..."

I imitated low groans getting gradually louder and louder and tailing off into a series of grunts. The group applauded.

"Thanks all! Before we go to dinner, listen up. Tomorrow, early breakfast at 6:30; come prepared to go out most of the day. We'll take a packed lunch and go down to the big swamps further south."

The lions came around midnight. Well-fed and curled in my comfortable bed, I snapped awake at the sound of padded feet thudding softly past my tent, and a soft rolling growl. Through the window screen, I saw a lioness crouching on the path in front of my tent. A male followed and mounted her. She gave deep growls, he a high snarl as she twisted around and slapped him. I heard the sound of tent zippers. Then Ray's loud whisper: "Has David got company?"

Marilyn: "Those aren't people! Jesus, Ray, they're lions! Right. Next. To. Our. Tent."

I said: "Hey, guys, isn't this great! They won't bother you. They're only interested in each other."

The lions, now lying side by side, seemed uninterested in anything.

Ray: "This is incredible! These tents are thin, though. Are we safe?"

Me: "Never lost anyone—yet."

Marilyn: "Somehow, that doesn't reassure me. And from your talk, we can expect...*that*...to happen every 15 minutes?"

"Give or take. Just hush and watch the fun!"

About ten minutes later, I heard roars from down the valley, the pride we'd seen by the bridge. The female sat up and gave a loud groan. Then the male began, too, lying down.

"Here we go." I said, "This is the Real African Experience."

There is nothing that can prepare people for the sound of two healthy lions

Mating lions

roaring 10 feet away. More zippers opened—or closed—and I heard more excited voices.

The lioness stood, arched her back, stretched, flicked her tail, and strode away, the male a pace behind her.

Day 3. Tarangire

At dawn, my little band straggled in for breakfast. Some red-eyed people were already there, gulping coffee; they hadn't slept a wink since the lions' visit. Others overslept, and I sent someone to rouse them.

Annette grinned, "Good morning. You guys had an exciting night down at your end. Those lions must have roared for at least an hour!"

I joked, "Thanks for arranging that! What a perfect introduction to Tarangire."

Annette lowered her voice. "Mohamed, our room steward, told me that your single lady—Barbara? —threw every loose item out of her tent, apparently trying to scare away the lions. I hope she's OK."

"Oh, dear. Another Lariam loony?" I said, referring to the psychotropic side effects of a common antimalarial. "I'll have to watch her. I can't believe doctors in the US keep prescribing that stuff."

"Don't we know it!" agreed Annette. "Anyhow, you'll have a nice cool drive to Silale today. Your lunch baskets are ready at the bar."

After breakfast, I herded my group to the cars. Adam was excited about the lions, which had passed the drivers' quarters on their way to us. He turned back to the group.

"Swahili lesson—*Habari ya leo?*" *What news of today?*

"*Nzuri!*" said Peter. *Good.* "But no sleep! Big Simba—roaring! Unnhhh! *Unnhhh!*"

Everyone laughed.

Our route went south along the river again, where baboons were descending from their roosts in tall palm trees. We turned east and gradually crossed over a

Elephants in Silalei swamp, Tarangire

low wooded ridge, stopping to admire two klipspringers, little antelopes standing very still on a rock as though carved from the same gray granite. As we came down towards the edge of Silale Swamp, three ostriches flounced away from the road. Silale was a tranquil scene, with beautiful umbrella acacias bordering the vast grassy swamp. We saw elephants drinking from pools, or rubbing themselves on tall rust-red termite mounds and themselves becoming rusty red.

"What are all those dots out in the swamp?" asked Peter, "are they elephants, too?"

"You're right," I told him, "at least 100. Green grass is their very favorite food."

Safari cars are like seagulls—or vultures. They gather where there is something good, and as more see the gathering, so the crowd grows. We spotted a gleaming huddle of cars in the distance around a big acacia, so Adam picked up speed. "It's either a rope or spots above," he told me in Swahili. This would make no sense to most people, and that was his intent. Some visitors knew the Swahili names of animals, so when exchanging news, the guides used a Swahili code-name for each animal. Thus: mustache = lion, ears = elephant, horn = rhino, water pig = hippo, scissors = crocodile, spots below = cheetah, spots above = leopard.

We joined the group under the tree. Adam soon spotted a "rope," a giant python coiled in the branches, digesting a belly full of fish. That was the day's most unusual sighting. We ate our picnic on some benches near the road and enjoyed watching giraffes, buffalo, impalas, and mongooses on the way back to the lodge and its refreshing swimming pool.

Day 4. Tarangire to Ngorongoro Crater

After breakfast, we loaded up and headed north along the floor of the Rift Valley, then turned northwest onto a broad, corrugated gravel road that would take us to Lake Manyara, Ngorongoro, and Serengeti. I rode with Sakala, who swerved side to side to avoid Adam's dust cloud and stones, and oncoming vehicles.

Soon we could see the western cliff of the rift rising 2,000 feet above the floor. We approached through farms and tall trees on the outskirts of a pretty little town. "What's this place?" asked Jane.

"Mto-wa-Mbu" said Sakala. "That means, 'river of mosquitos.'"

"You can see why," I said, "there is so much groundwater. Look, there are rice paddies there on the left."

Sakala continued, "One safari, I come here with a Japanese group. I say that name and *bam! bam! bam!* they closing all windows, then *pssst! pssst!* they spraying Doom! I said, no problem. Mosquitoes only biting at night!"

We laughed. Barbara asked, "Has either of you ever had malaria?"

Sakala and I simultaneously said yes.

Did we take anti-malaria medication? "No, people who live here don't use it," I added, "The drugs can have bad side effects. Instead, we learn what malaria feels like, and if it starts, we zap it with a drug that cures it in a day or two."

Sakala nodded. "In my village, we using a certain root. Boil in water, make tea, malaria finish very fast."

"Sounds risky," said Barbara, "but I wonder if that drug is affecting me. I'm not sleeping well."

"Maybe try doing without for the next few days," I suggested. "Where we're going, there's very little risk of malaria. See if you feel better."

A riot of color distracted us. Under the scarlet canopy of a flamboyant tree, women in brilliant kangas sat behind plastic bowls of green, yellow, and red bananas. Adam had stopped, so we stopped too. I bought some bunches of sweet red bananas and passed them around.

"Lunch at the usual place?" asked Sakala. I nodded, yes.

We ate in a picnic area just outside the gate of Lake Manyara National Park, with the strong smell of guano and a commotion of birds in the tops of tall forest trees above us. Sakala fended off inquisitive blue monkeys. This year the lake was full of water and fish, so the yellow-billed storks and pelicans were nesting in hundreds. After lunch, we ascended the steep rift road through bush and baobab trees, getting an aerial view of the big white birds sprinkled densely on the forest canopy like flowers. South of us, the lake reached along the foot of the cliff to the horizon. Above the rift, we drove through hilly farmlands, the fields green with maize, beans, and pigeon peas.

We sped on, past the junction of the Horrid Road, to the entry gate of the Ngorongoro Conservation Area (NCA). The drivers went to pay entry fees, and most of us headed for the toilets. I checked that our windows were closed and went over to Adam's car.

"Good idea to shut windows," I said, "Baboons are around."

As I walked away, I heard a commotion and a woman's scream. I turned just as a hefty male baboon galloped three-legged past me into the undergrowth, a box of cookies under one arm. Back at the car, Barbara wailed.

"Sorry! I was holding the door open for Peter when that *thug* of a baboon pushed past me, grabbed the cookies, and fled. He was so fast!" I laughed and reassured her that these baboons were cunning and lived by stealing snacks from an endless supply of naïve fools, though I used more diplomatic words.

From the entry gate, we ascended a steep winding road towards the crater rim. We had to squeeze past a heavy supply truck, supported on primitive wooden jacks with a rear wheel completely removed. A youth slept beneath it in the shade, surrounded by leaked oil and several days' worth of chewed-up sugarcane. The situation was under control; there was nothing we could do. Everything that anyone ate, used, or built in Ngorongoro and Serengeti had to come in trucks like this, up this road; many couldn't quite manage.

Soon we emerged on an open ridge that on our left fell steeply away to the farms of Oldeani, and on our right …

"Wow!" exclaimed everyone, glimpsing the serene green bowl of Ngorongoro

Crater. Our two cars parked, and we stepped out to gape at this new wonder. The air was cold.

"I had no idea it was so big," said Peter.

"Yes," I said, "the floor is about 100 square miles, and it's 2,000 feet below us. We'll go in tomorrow. But use your binoculars and see if you can spot any animals now."

At first, it was hard to grasp the scale—there were almost no familiar objects on the crater floor. Then we spied thin brown roads with tiny white cars crawling along them.

I found two elephants. "See down there? They're grey things bigger than cars. Then that clump of black things by that swamp, almost as big as cars, they're buffalo. And the smaller scattered brown ones are wildebeest."

"I don't like that gun." Marilyn indicated a young, green-uniformed ranger off to one side, an AK-47 in his arms. He smiled and waved, pleased to be noticed. "What's that about?"

"Just keeping us safe," I replied blithely, "from big animals, you know."

Armed bandits had recently attacked and robbed an isolated tour group at this spot. But my group didn't need to know that.

We continued around the rim of the crater to reach the Ngorongoro Crater Lodge. This was the original lodge founded in 1937, a simple cluster of black log cabins with mossy, shingled roofs. On its lawns, several large black buffalos chewed the cud.

John exclaimed, "Hey! Water buffalos!"

"No, just buffalos—African or Cape. Much more dangerous than their domesticated Asian cousins."

As we disembarked at the reception hut, I told everyone, "Watch out for those old bulls. They look tame, but they're completely wild and can kill you. So, don't walk up to them, and if one is near your path, ask a staff member to escort you, or detour around it. Survivors, let's meet in the bar at 7 p.m."

Buffalo grazing at Ngorongoro Crater Lodge

The bar and restaurant were a long log and stone building with a panoramic view over the crater. Inside, there was a cheery wood fire, very welcome as the evening air was already chilly. Friendly red-shirted waiters served drinks. I gave the group some background to Ngorongoro, then announced the plan.

"Tomorrow morning, we're up early! Breakfast here at 6 a.m. and then we head out with a picnic lunch. It takes about an hour to drive down there, and the best wildlife viewing is at dawn."

There were general groans and wails at the prospect of so early a start, but Peter said, "Get over it. We're going to see the Big Game in the Super Bowl!" More groans.

Over dinner, we entertained each other, making up stories about the other lodge guests. By doing so, we probably learned more about one another than about the other visitors. Eventually, we dispersed to our cabins, tired but happy with the trip so far. The best part was yet to come.

Day 5. Ngorongoro Crater

We had a peaceful night, save for the barking of zebras grazing and distant lion roars floating up from the crater floor. First light filtered through a light fog. Guards escorted us past dark animal silhouettes to the dining room. Fresh coffee, eggs, bacon, and toast.

"Were those wild dogs barking during the night?" asked Jane.

"No, wild dogs are scarce, and they don't bark. Those were zebras."

"There was a buffalo right outside our room!" exclaimed Danny. "We could hear him chomping. Dad wouldn't let us go out."

"Quite right. too!" I said. "Five minutes, and we're out of here."

We rattled around the crater rim and down the steep one-way road, our roof-hatches open. The blue wall of the caldera circled us, and the grass was green. Water glinted from ponds and the little lake on the crater floor. As the red road unfolded in front of us, we saw an abundance of animals—wildebeest, zebra and gazelles, warthogs so accustomed to cars that they wouldn't budge from their resting spots in the road. There were herds of buffalo, pools of splashing hippos, vast tuskers, a couple of distant rhinos, eland, hartebeest, waterbuck, jackals, hyenas, lions, and even a splendid pair of male cheetahs.

We crossed the small Munge Stream and drove up a low hill on the crater's north side.

"Here you can see the ruins of Adolf Siedentopf's farmhouse," I told everyone, standing up through the roof hatch. "He was a German who settled here until WWI. And this place was also my introduction to African wildlife, 24 years ago. In that bend of the stream, I spent a week camping with Jane Goodall and her husband Hugo van Lawick and their little son, Grub. She was studying hyenas, and he was filming them. I was on my way to Gombe to work on Jane's chimp project for two years. You can see how I got hooked on this gorgeous place full of wildlife."

CHAPTER 4: A TYPICAL TOURIST SAFARI

Kori Bustard

I don't like name-dropping on safari, but coming here always reminds me of all that I owe Jane.

We met a grey, goose-sized bird pacing on long legs slowly through the grass. It cocked its crested head to one side, then stabbed a large grasshopper. Sigh. I knew what was coming:

Me: "That's a kori bustard at two o'clock!" Behind me, a motor-drive went *click-clickclickclickclickclickclick.*

John: (who had just taken 37 photos of it) "Hey, what have we here?"

Me: "It's a kori bustard. It's the—"

Betty: "My goodness, what a big bird. Is that a secretary bird?"

Me: "No, it's a kori bustard. You can recognize it by—"

John: "It's a what?"

Danny: "Dad, he said it's a coreybuster"

Me: "Bustard. B-U-S-T-A-R-D. They're a family of—"

John: "OK, gory buzzard. But it doesn't look like a scavenger …"

Jane: "Just look at the size of that thing. That one of those crowned storks?"

Me: "No, it's a Kori. Bustard. Rhymes with mustard. Ah, look, Adam's found it in the book for you!"

Lisa (removing her earphones): "What are we stopping for here?"

Sakala was watching me from his car, one hand covering a grin. I mimed tearing out my hair in large handfuls, shooting myself, and collapsing on the roof.

Whoever named African birds so awkwardly? Every trip, we must explain kori bustards, augur buzzards, crowned hornbills and ground hornbills, lilac-breasted rollers, lappet-faced vultures, and the like. To add to the confusion, I had created an "unreliable field guide" to East African birds, with cartoons illustrating the more peculiar names; we sold it in the tourist gift shops.

At midday, we reached Ngoitokitok Springs, the main picnic site. We saw a beautiful blue pool bordered by a swamp of tall reeds. Rafts of hippos floated on calm water. An elephant browsed among fever trees on the far side. A small building with a big plastic water tank promised relief from a pressing need. "An army marches on its stomach," I observed, "but a tour group marches on its bladder! When you're done, walk over to the cars; we'll be over there." I pointed to the far end of a crescent of about 50 cars, at the east end of the pool. Sleek dark raptors circled and swooped over them. Kites. Trouble.

The drivers spread Maasai blankets on the ground, laid out our lunch boxes, and a cooler full of water and soft drinks. We opened the boxes. Birds appeared like magic.

"The blue ones are superb starlings, the yellow ones are Speke's weavers—Betty, we're not allowed to feed them—oh, and here comes a guineafowl—but watch out for those dark brown kites overhead, they want your meat! So, guard your lunchbox!"

"That group has a table and chairs," observed John. "Why can't we?"

"They're with Uppercrustleys," I replied, "and they are paying more than twice as much for their safari as you are."

"I don't like chicken," said Lisa. "Anyone want it?"

"I'll take it," said Danny, reaching across for the drumstick in her hand, when...

Whoosh! A kite swooped out of nowhere on wings five feet wide, grabbed the chicken in its claws, and soared away. Lisa screamed, then looked embarrassed. But our surprise was quickly eclipsed by a more significant excitement—about six tons of it. The elephant we'd seen earlier had decided to walk round to the other side of the pool. His path took him through the picnic crowd. The people closest to him were shouting and laughing, gathering up their lunches and cameras, and scrambling into their cars. The commotion spread ahead from the plodding tusker,

Black kite stealing a chicken leg

and soon we, too, were seeking safety. The magnificent bull strode past us, ears flapping, massive appendages swaying.

Jane gasped, "Lookit. Is that his...Oh. My. God!"

We finished our lunches in the cars, while the "two-trunked" elephant drank slowly and deeply from the pool and then wandered out of sight.

After lunch, we continued to Lerai Forest, a stand of fine old fever trees. There were more elephants, buffalo, waterbuck, and lugubrious marabou storks. A steep and rocky road led us out of the crater and back to the rim.

On our way to the lodge, we stopped at the new Maasai boma that the NCA had set up to allow visitors to take photographs and buy spears and beaded mementos. It was a tall stockade of split juniper poles enclosing a circle of rounded huts of bent sticks plastered with mud and cow dung, and a bare, muddy arena in the center. There were women in blue or purple cloaks, with shaved heads and flat beaded collars. The men wore red or maroon cloaks; warriors had beaded belts with a short sword in its scabbard and a hardwood club, and each carried a steel spear as tall as himself. Some had their hair in long braids. The elders had short hair or shaved heads and just carried a walking stick. The tallest elder greeted us in proper English.

"Welcome to Maasai cultural boma! My name is Taté, and I am your guide."

We negotiated a photo fee of one dollar each, a terrific deal.

Taté gave a short introduction to the project and told us that it had raised enough money to buy livestock and food and pay for school fees for many youngsters. The day was bright and lovely, and my group wandered in and out of huts and around the boma, navigating between the cowpats and flapping uselessly at the affectionate flies. Eager young warriors quickly surrounded Lisa, sweet sixteen in short shorts, her face as red as their cloaks. Betty found a necklace, and Peter bought a spear. The group could check off the Colorful Tribal Encounter from their list.

Back at the lodge, no one stayed up late except me, unwinding by the fire as I liked to do, alone. A young woman tour leader joined me, in shawl and jangling bangles, desperate to share safari angst with a Real Person. I soon escaped and took my chances with the grazing buffalo.

Day 6. Olduvai Gorge–Serengeti Plains–Lake Ndutu

After breakfast, we headed for Serengeti. The road down to the plains was steep. We passed a crumpled Land Cruiser, wheels up, that had flipped on a sharp bend—not an unusual sight. At the foot of the slope, Maasai women filled plastic jugs from a muddy pool where a rocky stream crossed the road. The trees thinned out, and we were on the short-grass plains. Thomson's gazelles were everywhere, wagging their black tails and playing chicken with the cars as we rattled along.

We turned off at the sign to Oldupai Archaeological Site. More miles of spine-jarring washboard brought us to the Visitor Centre, a small, two-room, tin-roofed museum and a couple of breezy thatched shelters overlooking the famous gorge. I pointed out the somewhat dire toilets and told everyone to gather in the shelter by the museum. Meanwhile, I signed us in and found the guide to orient us to the gorge's history.

He delivered a reliable product, the same talk we'd heard many times before. I almost knew it by heart, including his accent and eccentric syntax. First, he thumped the table with a tired-looking piece of a succulent plant about a yard long. "Oldupai Gorge is known by Maasai name of this plant. Germann wrongly called it Olduvai. Dr. Louis and his wife Mary Leakey, who are the owthors of the gorge, were not the fest to discarver this weldifamous site. In 1911, an entomologisty from Germanni was chasing barterflies on foot, from Ngorongoro across the Serengeti Plain, when he found himself, accidentally, down into the gorge. While looking here, and there, he discarvered large quantities of fossil bones...."

He droned on. I would have to recap the history more understandably later. Our guide escorted us down into the gorge. Modern time machines, our Land Cruisers lurched down through two million years of history. We reached the bottom at lava level, laid down nearly two million years ago, and crossed a stream. At the "Zinj site," a concrete plinth stood in a small quarry. Here, a small, cracked marble plaque recorded Mary Leakey's discovery of "Zinjanthropus" in 1959. Finding the famous skull at the foot of a slope, the Leakeys' team had cut into the slope in search of more of the skeleton. They didn't find any, but removed and sifted hundreds of tons of soil in the process. Surrounding the plaque was a wonderful collection of fossil fragments and stone artifacts we could handle and discuss.

We returned to the top and spent half an hour at the museum. Predictably, the keen ones tried to absorb all the very concentrated information, and the others bought souvenirs from the array of Maasai beadwork outside. Meanwhile, I sold our Ngorongoro guidebooks to the lad who sold souvenirs. We ate our Crater Lodge picnic lunch in the shelter; I ate everything in my box, but my wasteful Americans left a lot of food uneaten. From the bushes, two Maasai teenage boys watched us expectantly. They looked like specters in black cloaks, with intricate white masks painted on their faces. One wore two black ostrich feathers in his headband.

"What's with those guys?" asked Jane. "We haven't seen any like that."

"They're Maasai boys who have recently been circumcised," I said. "This only happens every few years, and all the teenage boys in the neighborhood are 'initiated' together. It's a coming-of-age ritual and a huge deal, because the operation is excruciating, yet the boy must not flinch in any way if he is to be seen as a real man. Afterward, they dress like this for about two months before taking on regular warrior uniform."

"Wow!" said Peter. "If I had my willie whacked, I'd have a white face too! Can we take photos? Of their costume, I mean."

Everyone laughed. "That's why they're here," I said.

Adam negotiated quickly in Swahili, then announced— "Ten dollars for them

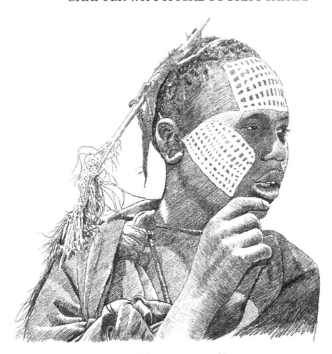
Maasai circumcised boy

to share, and everyone can take photos." People scrambled for cash and cameras. We got our pictures, and the boys got their money, plus all our picnic leavings when we departed.

From the visitor center, we crossed miles of open short-grass plain, then entered a brushy valley. Water flowing across the road stopped us.

"Where's all of this water coming from?" asked John.

"It rained on the plains last night, maybe around Ndutu," I said.

"Will we get through this?"

"Hamna shida, *no problem*," said Adam. "If the water is not touching that big rock, we can pass. Please shut windows." He gunned the engine and plowed into the muddy torrent while water sprayed over the hood and windscreen. "Tanzania car wash," he said happily, as we emerged dripping.

Beyond Olduvai, we saw our first wildebeest, coming in long lines from the east, thundering across the road like freight trains, black manes and tails flying.

"Here is migration!" said Adam. "They see rain at Ndutu, and they go there. This will be very good for us."

On the dead flat plains, Danny pointed to a large body of water, reflecting a small hill.

"Is that a lake or just water from the rain?"

I smiled. "It isn't there at all. It's a mirage. Watch, it's disappearing already."

"Wow, cool!"

We turned off the rocky road on the Ndutu track, just parallel tire tracks heading into infinity. We paused to take in the immensity of the herd that surrounded us. Thousands of wildebeest, mostly with heads down, grazed contentedly or sat ruminating. Here and there, a bull cantered with head up, grunting loudly at neighboring bulls. Small tan calves stood or suckled or frolicked around their mothers. Bands of freshly painted zebras moved among the dark wildebeest.

A mud-spattered Land Rover pulled up beside us. In the back stood a couple in matching new Banana Republic outfits. Their young driver started chatting with Adam in Swahili ("News?" "Nothing." "Me, neither."), while Mr. Banana accosted me.

"Hi, how y'all doing?"

"Great! Can we help you?"

"Well, yeah...have you seen the Migration?" he asked.

"Fabulous isn't it!" I said.

"No, I mean, where is it?"

I looked at my companions, and they looked at me. Tell 'em, David. I gestured at about 100,000 wildebeest and 5,000 zebra.

"Well...there are a lot of animals right here. You want more?"

Ms. Banana shook his head, frowning, "No, no, but these are just standing around eating," he complained, "We were promised we'd see the Migration. Like on TV. Herds stampeding, crossing the river, swimming with alligators, like that."

"Ah. Well, you won't see any of that here. What you see here is the most important part of the migration. The animals want to stay as long as possible on this plain, eating good grass, giving birth, and rearing their babies. Then in two or three months, these plains dry up, and they all head back north. That's when you get the river crossings and the, ah, *crocodiles*, a hundred miles north of here. Your guide should know all this."

Mr. Banana looked disappointed. "Well, he did tell us this is the Migration. Guess we expected something a lot bigger. Thanks anyway."

"If you're staying in Ndutu area," I said, "you may get lucky and see a herd cross one of those lakes. Good luck!" They nodded and prepared to leave.

Ray called out, "You should have *our* guides!"

Marilyn tried to shush him. "Raymond! That isn't nice!"

Adam shook his head and chuckled.

We drove through scattered herds for miles, then we entered an acacia woodland with groups of impala and scattered giraffe. We paused to watch elephants bathing in a pool as we crossed between Lake Masek and Lake Ndutu, then headed up to the lodge. Ndutu Safari Lodge was a cluster of thatched buildings shaded by beautiful acacias. Jeannette and I had been coming here for 20 years. We knew the managers, owners, and staff well. Sometimes we even helped out as caretakers.

While I was checking in, I sent the group into the open-sided lounge, to gaze over Lake Ndutu and sip coffee or a soda. "Karibu, mzee!" said our friend Aadje, the part-owner of the lodge. "You've come at a good time. Gnus everywhere, the Marsh pride has little cubs, there are a lot of cheetah families, and this evening you'll meet our fireside dikdiks. When you've been to your rooms and freshened up, lunch is on the table."

That afternoon we had a marvelous drive in the woodlands, then gathered in the bar in the evening. Genets with long ringed tails prowled the rafters like slender cats, looking for insects drawn to the lights. A campfire lured us into taking our drinks out to the fire circle. An old waiter brought out bowls of popcorn and nuts. As the sunset glow faded behind flat-topped acacias, a million stars appeared above us. Aadje came with a tour-leader friend of ours and his four clients, staying in the lodge in between camping locations.

"Hey, Dave, how's it going?"

"Pretty well, Nige," I said, "How about you?"

"He means fantastic!" enthused Peter. "You wouldn't believe what we've seen!"

Uh-oh, I thought. Never start a pissing match with Nigel.

"Oh, we had quite a good morning here, I guess," Nigel said, deadpan. "We were with this leopard and her cub for about an hour, right near here, climbing trees, playing, and then the mother saw some gnus with calves and went stalking. She vanished completely for ten minutes in weeds only a foot high. Then she popped up, grabbed a calf, and killed it right in front of us! But, before she got a bite to eat, two young male lions came trotting across the gulley. Now that mama wasn't going to lose her lunch. She grabbed the calf by the neck and took it *straight* up a tree, 20 feet up, cub following. Nothing the lions could do, but they settled down to wait under the tree. The news spread, and soon there were 30 bloody cars there, so of course, we buggered off and watched the wildies swimming across Lake Masek."

All adoring eyes were on Nigel. He was The Man.

"That must have been fantastic," said John enviously. "David hasn't found us a leopard yet."

Never mix safari groups, I thought bitterly.

"Damn, Nige," I said, "you've always been so lucky with leopards. This afternoon we went the other way—south. All we saw was a wildebeest birth, the whole thing, from contractions to a wobbly baby and the imprinting process." I paused to let the fantastic scenario play out in everyone's head. "Mind you; we did see a pangolin..."

"*Get away!*" said Nigel, "You didn't! You lucky sod."

"What's a payngolin?" asked a woman in Nigel's group.

He explained, "It's a mammal covered with scales that can roll itself in a ball, sort of a scaly anteater. But very rare; I've never seen one."

"Oh, like an armadillo. We have 'em in Tayxas"

"Davidi, everyone," interrupted Aadje, "look, Mr. and Mrs. Dikdik are here."

In the firelight, we could see two dainty little antelopes beside her director chair, delicately nibbling at some spilt popcorn.

The waiter brought more beers, a hyena whooped in the distance, and we continued to weave stories into the magic tapestry of the Ndutu night.

Cape Pangolin

Day 7. Ndutu Woodlands and Serengeti Plains

We headed out at first light to look for the leopards. Sakala had precise instructions where they were, and they rewarded us with some quality time.

Throughout the day, we cruised from spectacle to spectacle. Sometimes the "small game" is as memorable as the "big game."

At the bar that night, Rob and Lana joined us, a young couple studying dung beetles on the Serengeti plains. Rob was telling us about the 100 different species they had identified, each with its strategy for eating poop, when the biggest of them all, like a shiny brown golf ball, came droning in, smacked into a lamp, and fell—on Lisa's head. Oh, dung! She leaped up, knocked over her soda, flailed at her hair, and screamed, "*Ewww, getitoffgetitoffme!*"

"You're OK, I've got it!" Lana calmly caught the beetle in her hands and gave it to me. "*Heliocopris*," she said. "Isn't he a beauty?" Lisa crept back, embarrassed. If Lana, a girl only a few years older, could handle this monster, it must be cool. "Sorry!" she said, "but bugs scare me."

I let the giant beetle clamber over my hands to calm it down. "He can't bite, he's harmless—just clumsy and prickly. Here, touch his back, I'll hold him still."

Gingerly, she dabbed at him with a finger and survived.

"He's so strong," I said, "that I can't hold him if he wants to escape." I wrapped my fist around the beetle, but he forced his way out, using head and arms like a wedge.

Lisa steeled herself. "OK...let it walk on my hand. But don't let it fly!"

I placed it on her hand, then caught it as it tumbled off.

Lisa beamed, "I did it! I did it!" And everyone cheered.

Day 8. Ndutu–Serengeti–Gibb's Farm

We had to return to Arusha the way we came. On the journey back, we stopped at a special place for the night. We took a muddy track going up towards the forest edge, approaching Gibb's Farm through beautifully tended coffee fields. We arrived at a lovely garden with little tables and chairs on a grassy slope shaded by lofty trees. This was "home from home" for Jeannette and me. Margaret and Per were a source of strength and inspiration to us. Margaret rose from one of the tables to take my hand.

"Hello, David! Welcome to you and your guests." She beamed at the tired group straggling up the lawn. "Do have some lunch," she said, waving at the open doors leading to the lounge and dining rooms.

After an excellent lunch and a rest, I gathered the more energetic members for a walk in the forest. Our guide was Gie, a local Iraqw man who had been an anti-poaching ranger with the NCA. A hard man with a hawk-like face but a twinkle in his eye, he had retired from that dangerous life. Now he tended Gibb's Farm tree nursery, growing tree seedlings and persuading villagers to plant them. He led us along a contoured trail through two valleys to a stream, where elephants had excavated a great cave in the riverbank in search of clay minerals to aid their digestion. We hopped across the creek and entered the cave and marveled at all the tuskmarks in its wall. From the cave, we followed the stream down to where it plunged 100 feet over a cliff on its way to the farms below. The breeze blew refreshing spray in our faces.

In the evening, I set up a circle of chairs on the lawn and invited all my group and Sakala and Adam for a farewell drink.

"Since I met you all at the airport—was it only two months ago? (laughter) —we've been through a lot together and shared amazing wildlife experiences. I want to especially thank Sakala and Adam for all their tireless, cheerful work, and brilliant animal spotting."

There was applause, and Ray handed each guide an envelope: "From us all, *asante sana!*" They looked delighted and shook his hand, stowing away their envelopes to be opened later.

"I'd like to thank you all too for being such a nice group to work with, and I'd love it if we could go around the circle, and everyone tells us your high point of the trip. I promise I won't make you hold hands and sing Kumbaya!"

Predictably, people raved about the roaring lions, the elephant at the picnic, the vast herds of the Serengeti, and our wonderful encounters with predators. But I listened for what they said about themselves or didn't. And as always, there were a few surprises.

Lisa: "I didn't think I could last ten days without TV or phone, but I've done good! And look, no earphones! You're gonna laugh, but a big thing was when David made me hold that dung beetle. I was always terrified of bugs! It was so cool meeting the people who study them. I think when I start college, I might change my major to biology." (Applause).

Danny: "The animals were amazing, especially the leopards. But I really enjoyed meeting the Maasai and talking about soccer with Adam. He plays striker, just like me."

John: "This has been a great trip for all of us. The kids have gotten into it. And Dave, you just kept hitting us with one surprise after another. I guess what I'll remember is that incredible 360º view of the wildebeest on the plains—and those dorks looking for the migration. Oh, and you might have noticed I'm a picky eater, but the food has been delicious. I tried some new things, like custard apples and passion fruit, and liked them!"

Barbara: "I feel like I missed the first three days because of that dreadful drug. Since I stopped it, I'm back to normal. Traveling with you all has been a pleasure and getting to know a new friend!" She blushed and reached out for Peter's hand. (Well, well!)

Peter: "Exceeded my expectations in every way! That's the short answer. I could have gone home happy after the first day, but every day kept getting better. And your sneaky British wit, David; your David Attenborough commentary about the other diners in the Crater Lodge was a gem!"

Betty: "No one's mentioned the birds! I came for the big cats, but the birds blew me away. I bought a bird book at Tarangire and started checking them off. With help from David, Adam, and Sakala, I've got 162 species for the trip."

Jane: "Betty and I travel a lot together, mostly on historical or cultural trips, but this has been one of the best. I'm in love with elephants! When we get home, I want to see what I can do to support protecting them."

Ray: "We've been to Africa before, but South Africa; it's more developed, sure, but it can't hold a candle to Tanzania in terms of the wildlife we've seen. I feel like we've lived in a National Geographic special for the past nine days. I liked the way we did it, just moseying along and letting things happen. As an airline pilot, I've always needed precise schedules, but now that I'm retired, I'm starting to enjoy 'going with the flow.'"

Marilyn: "Everyone's already voiced my thoughts, but I loved the warmth and friendliness of all the Tanzanians that we met. In the West, so much of our news of Africa is negative—wars, corruption, famine—so it's great to see a country trying to do things right, protecting the wildlife and educating kids and building for the future."

Sakala and Adam smiled inscrutably. Sakala stood up. I was glad the old pro would give the final words.

"Thank you, everyone, thank you, Mzee David. My colleague and I, we thank you for coming on Hatari Safari. For seeing our wildlife, our country. We hope you are coming back again soon. *Karibuni Tanzania, asanteni sana!*"

Day 9. Karatu to Arusha
After a scrumptious breakfast, including fresh-baked pumpkin nut muffins, we loaded up and faced the bumps and dust again. We aimed for a late lunch, then a chance to shower and repack at Fisi Lodge.

Fast forward—we reached Arusha safely, stopped to buy souvenirs, had a forgettable lunch, used our dayrooms, and went with Sakala and Adam to the airport. Kilimanjaro dutifully emerged from the clouds just before sunset. At the terminal, my nine new friends, no longer khaki clones but unique individuals, parted with hugs and tears and promises to keep in touch. Ray, as the group elder, handed me the ritual envelope.

"So that's how it was," I told Jeannette over a beer at Fisi Lodge, "I've left out such a lot, but it was a good safari, and everyone left happy."

"Thanks for the *ripoti*, my love. I knew you'd have a good time; you usually do."

A pause. Jeanette looked expectant.

"Oh yes, hmm, how was your horse safari?" I asked.

She laughed. "We had an exciting time. About 12 of us, including a diplomat with a mistress; a lone lady USAID worker looking for action; Mafalda, that strange old German horse trainer; two teenage girls just learning to ride; and Paul, our Irish fish farmer friend. At the end, the police arrested us! The army told them we had illegally photographed a communications tower. It was just part of the background in our pictures, but they took all our film and trashed it."

She paused; I knew there was more to the story. She went on.

"The horses were skittish retired racehorses, not at all suitable for trekking in the bush. We rode around the north side of Mt. Meru, down into the Great Rift Valley. The scenery was amazing, but the horses kept acting up. At one point, they all decided to race across the thorn bush. My horse leaped over a young acacia, landing me on my back among the three-inch-long thorns!"

"What? Are you OK?"

She chuckled.

"Well, I'll show you the damage later. The thorns went into my back, butt, and head. Paul and the others poured whiskey on me and pried out thorns at our camp that night and more at the lodge the next day. But I can still feel them. I'll need your help to get the rest." She laughed and I joined in. *Safari njema*, indeed.

And so, we spent a prickly night together and left for home the next morning. We were cheerful, still in safari mode, ready to cope with whatever disasters or delights awaited us in Mangola.

Whistling Thorn - this tree not only has two-inch thorns, it provides free housing for biting ants who help defend it.

CHAPTER 5
KIBUYU PARTNERS
OUR BUSINESS RELATIONSHIP AND A BATTLE WITH BUREAUCRACY: 1992

Boating through the papyrus at Mikwajuni

David shoved off from our makeshift dock on the edge of the stream. I almost fell in as he pushed away with a strong pull of the paddle. "Hey, wait for me," I called as I lunged at the end of the boat.

The punt was a long flat-bottomed boat made for slow streams. David built it by hand, lovingly curving the sides and creating a platform on both ends. Gratefully, I landed on the platform and lowered myself onto one of the damp cushions in the bottom of the boat. Taking the exaggerated pose of a leisured lady, I settled my bonnet on my head and admired David. We'd enjoyed punting during our courtship in England, and I still loved sliding along with my beau stroking the water. I chuckled at the image.

David paddled upstream to a place where the papyrus on both shores hid us from view. We had prepared to fight it out. We had paper and pens, a bottle opener for two beers, and a big sharp knife. Our aim—to agree on a logo and a name for our business. We'd been married for twenty years and had been "in business" together for virtually all that time. Our business and aims in life included learning, educating, and sharing. We wanted to produce things of beauty and importance within our capabilities. To do that, we had to learn to work together.

Two people with disparate styles meant we didn't find it easy to work as partners. I'm the kind who likes clear plans, pacing, and meeting deadlines. David is much more relaxed about making plans; he considers deadlines as flexible as bungee cords.

And the big sharp knife? Oh, it was for cutting the papyrus stems blocking the river channel. The monkeys and baboons knocked them down to make bridges. Today we needed to work on a bridge between our two minds. For over ten years, we had designed different logos and different letterheads. We could not agree about any. Our little business needed to use stationery that required an official-looking stamp. The government wanted us to register our correct business name so they could tax us properly. Because we'd never reached an agreement, our lawyer had registered us as "David Bygott & Co."

Yeah, well. Was ol' "Co." pleased with being an afterthought, an addendum? Not on yer life, mate. But what could I do about it? Change it, that's what. But I couldn't come up with anything better. For the years that I had been the "& Co.," I endured people telling me how great David Bygott's work was, when it was at least partly or wholly mine. I was not humble enough to let my spouse and partner get all the credit. I wanted some, too. Besides my creative input, I put in time, lots of it. I also had the emotionally draining, gritting-the-teeth task of pushing and pulling, enticing, and cajoling my partner on every shared project.

David and I have a "rule" about conflicts: whoever cares most about doing something must take responsibility for getting it done. We use a scale of one to ten to determine the intensity of any concern. In this case, I was eight on the ten-point scale; David managed a balanced five. Hence, it was my responsibility to force us to find the right company name together. Our day in the boat was the ultimatum: we had to agree on a name. An added demand was deciding on a symbol, a logo for our business.

We parked the boat in a spot surrounded by a screen of papyrus and listened carefully to the swamp papyrus gossip. Voices let us determine if the illegal distillers were busy in the swamp, snorts told us if hippos lurked in a pool upstream, munching revealed goats or cattle browsing alongside the river, and barks and rustling prepared us to be descended upon by monkeys or baboons. Lucky for us, all seemed quiet, papyrus heads simply muttering in a soft breeze. Sitting in the dappled shade of an overarching fig tree, we engaged our imaginations to search for an image that would represent our business. We designed logos of birds, mammals, flowers, trees, and humble household objects. We agreed on none.

At last, the "aha" moment came: the idea of gourds. Gourds were ubiquitous, in every household, often beautifully carved and decorated. They had multiple uses, for carrying water, milk, beer, and honey as well as seeds and foods.

CHAPTER 5: KIBUYU PARTNERS

Our Kibuyu logo

We drew gourds in various ways, sizes, and positions. And yes, we reached accord on the gourd. Our favorite sketches showed small, rounded gourds (me) and longish, sturdy gourds (David). David added a smart touch by putting paintbrushes, pens, and pencils sticking out of the little gourd.

Delighted with ourselves, we opened the beers, and congratulated each other. Then we just lay down on the cushions and gazed around our outdoor office. Looking up into the leaves of the fig tree above us, we saw the long russet tail of a paradise flycatcher. We heard distant shrieks of hadada ibis and the nearer chatter of lovebirds. They resembled animated green-yellow-red candies clambering among the figs. This was one of those precious moments when we felt in harmony with one another and our precious wild world.

The lovebirds rose like ripe feathered figs and left the tree. We summoned ourselves out of bliss to the business of finding a name for our business. That was easy. Following the image of gourds, came the name, "kibuyu." Kibuyu is the Swahili word for gourd or calabash and gourds were a recognizable symbol for sharing.

So Kibuyu seemed perfect. I wanted it just that simple. But alas, simplicity is not a hallmark of bureaucracy. The complicated, entangling red tape of bureaucracy trapped us when, months later, we arrived in Dar es Salaam to re-register our business name.

Dar es Salaam is on the Indian Ocean coast of Tanzania, not far south of the equator. The city is a hot and humid place. Some Europeans irreverently call it "Dar es Sal-armpit." We arrived on a still and sweaty day and found a parking place near the business registration office. There was no shade, but plenty of trash and irritating boys who wanted to guard our car. David fended them off and prepared to defend the vehicle himself.

I put my mind into patience mode, a mental gear between neutral and low range, ready to be stoic or climb hills. I headed into the ugly building. All the offices on the ground floor seemed closed or empty of people. I climbed the stuffy stairs to the first open office. A large woman sat at a table, reading a newspaper.

"Where do I go to register a business?" I asked, after offering my most polite Swahili greetings.

She told me, "This is not the registration office."

That didn't answer my question. I switched into low gear.

"Please, can you tell me where I should go to register a company?" I tried once more.

The woman with the carefully braided hair just looked at me. I was in her power. She did not smile. She pointed back out her door and said, "Upstairs."

I could have used more direction but realized the danger of asking. Upstairs I went. That floor had a big open office. I entered, but no one looked up from his desk.

I greeted the first and friendliest-looking man. "Is this the business registration office?"

"No," came the minimalist reply.

"Please tell me where I should go."

He looked me over then pointed back out the door. "Upstairs."

I turned with lips tightened in determination and headed up another squalid

flight of stairs, avoiding the shiny, sticky patches. I reached a landing, then another dingy staircase to another floor. All the offices on that level had closed doors. After another flight and another landing, on what I reckoned might be the fourth or fifth floor, I found a large, dimly lighted open plan room, filled to the brim with sweating men. They had acquired the same hue as the piles of damp brown folders around them. I went through my greeting routine again. The man I'd accosted directed me to the back of the office.

In a far corner sat a man at a desk piled high with folders. His bald head faced down towards some papers. He did not look up as I approached. I had time to notice the small dirty window open behind him, letting in some moist air that smelled like moldy fish crates.

I stood and gathered my courage. The man lifted his head. I asked him, in what I hoped was good Swahili, about registering our business under a new name. He looked at me as if I wanted to change the name of the Catholic Church.

"Madam, why do you need to change a perfectly good name to something new?"

"I can't explain now, but we want to change the name of our business from David Bygott and Co. to Kibuyu."

He stared at me. "No."

"Why?" I asked in dismay.

"The name is too short. A business name should have more words."

I was stupid enough to ask, "Why?" again. David often reminded me not to ask "why" questions, especially not in offices. But the question popped out of my untamed mouth. I couldn't fathom why a name had to be longer.

He didn't answer but just said, "Kibuyu...what?"

I stood like a statue trying to think this out. Taking a deep breath, I looked around while the man went back to his papers. The stuffy room was like a crypt, the zombie workers moving slowly, the supplicants bewitched into immobility. Stacks and stacks of wrinkled and wrenched files covered each of the four desks, the one huge table, and many wall shelves. Sweating people pored over the papers. I stood in neutral mode, breathing shallowly to both calm myself and avoid the smells.

Our carefully selected and simple name wasn't acceptable? Kibuyu what?

While I dithered, trying to understand what I could add, the man disappeared. I was terrified. Would I ever get him back? So many others awaited his attention. I had climbed up many flights of stairs between landing after landing of gloom and stink in this old, derelict building with little air or light. I'd spent agonizing minutes to find the right room on the right floor. I was desperate to finish this visit and flee. What could I add to the name?

I mulled and masticated names. At last, my mental light bulb began to glow—Kibuyu Partners. It seemed fine to me and made it even more apparent that David Bygott had a partner: me. Relieved that I'd come up with an addition to Kibuyu, I wrote the name out on a piece of paper. I stood at the desk, waiting and sweating.

People frowned at me. I was taking up space, producing body heat. I went downstairs, searching for my bald man in one dim room after another. Dejected, and thoroughly apprehensive, I returned upstairs to his office, drops of sweat rolling down between my breasts and dribbling down my back. The overworked man was back in his seat.

I smiled fervently in hopes he would find my modified name acceptable. He grumbled as he read the extended title. All I could deduce from the grumble and harrumph was the name might be acceptable. He didn't look up but scribbled on my forms—Kibuyu and Partners. I had no idea why he added the "and" between Kibuyu Partners and wasn't about to ask. The tired-looking man raised his eyes and said, "Now, you must have your partner sign this form. It must have his signature." He handed me the forms.

My brain sweated as I hiked back down the million or so stairs to find said partner. Neither he nor the car was where I'd left them. I was close to panic. Then I spotted David and our car far down the street, sitting under a magnificent mango tree. Relieved to find him, we discussed the new name. He signed the form. I didn't dare to look into his face to see if he was hiding a smile or a smirk. Back up the many flights of stairs—why are they called flights, I wondered? Terraces of Dante's Purgatory seemed more apt. I trudged up and up. Back to the overheated room. The bald brown man accepted the signed form silently. He handed me a new form.

"Now, you must pay."

"Where?" said I, switching my brain to low range.

He gestured out the door. "Downstairs."

Down I went, relatively light-footed, to pay a tiny sum to a bored clerk behind a barricade. Back up the stairs to submit the form and my receipt to the overworked man in the overflowing office.

Hope rose in me as I handed him the papers, nearing the completion of this awful task. But he grumbled again. "Now, take the forms to the typists."

"What?" I gasped, betraying too much dismay. "Typists?" I added quickly, "Where might I find the typists?"

He pointed out the door, "Downstairs." Taking pity on me, he added, "Two flights then go right."

I finally found the room by knocking on each of the closed doors, then trying the handle. As I slowly opened a door, I saw two women. They sat at desks free of folders or forms, just a typewriter each with a stack of paper and carbon. I looked at them, one very tubby with a round passive face, the other thin and sour-looking. Fat One wore a tight, mustard-colored polyester blouse that made me feel hotter just to look at it. Thin One had oily hair, and a newspaper spread in front of her. I coughed politely and even more politely asked for their help.

"Good morning. I wonder if one of you could kindly type this form for me? We are changing the name of our business."

They looked at each other, then at me. "Hmmm," I could almost hear them thinking, "Here's a white woman begging for our attention. It is Saturday. We want to go home, not do what she wants."

"We are busy," announced Fat One. "We can't type any new forms today. We close the office soon. Come back on Monday."

"Oh no! Please, can you do it today? I could make it worth your while," I hinted.

No response. I waited. Fat One glanced at Thin One, and both shook their heads. I knew it was hopeless.

Back I went to the bald brown man bravely fighting the paper tidal wave. "No," he told me, "I cannot help you. I do not type. I have no typewriter. All the papers in this office have to be typed," he paused, "by those...typists."

I sensed just a trace of scorn in the last sentence and felt a wave of sympathy. He too must have been frustrated by the women downstairs. Putting my angst aside for a moment, I took a closer look at this man. He was an office worker, obviously not very happy about his work or where he had to do it. I thought back to the time David and I sat in the boat watching the birds in the fig trees, smelling the river, deciding about a business name in a wonderful wild habitat, our outdoor "office."

What a difference in our lifestyles—this hard-working man, poorly paid, cooped up, living in a crowded city, and we foreigners, maybe not well paid, but living in a relatively clean and comfortable rural home. How silly, how insignificant, and how privileged we were, compared to this fellow. My distress at not getting my chore done changed into something else. Empathy? Shame? Understanding? I made a promise to myself that thereafter, I would make a much bigger effort to listen to and help the myriad supplicants who came to my own "office" in Mangola.

Since I still stood there, not saying anything, he looked up at me. I wonder what he read in my pale face with his brown eyes.

"Come back on Monday," he said.

I shook my head automatically as though to shake away the thought. Tomorrow—Sunday—we'd hoped to be on our way home. The idea of hanging about any longer in hot, unpleasant Dar es Salaam appalled me. What could I do?

I pleaded. "Please, can't you help me get this typing done today?"

"No," he said definitively. "Come back Monday."

He bowed his shiny head to his paper pile.

I bowed mine and looked at my feet, trying to fight off unwanted tears. I knew that unless I tried harder now, there would be nothing I could do but face all those stairs and floors and faces on Monday. And I feared the typists would still be too busy to deal with my typing request. I realized I probably couldn't deal with it and would return to Mangola without our new name.

My fidgeting may have driven the bald man away because when I looked up, he'd disappeared again. I surveyed the room in desperation and decided to escape, new business name or not. As I moved towards the door, another supplicant entered the crowded room. He tried to push past me, and I backed into a stack of files on the table behind me. The folders fell in an avalanche to the floor. Mortified, I bent over to pick them up. Someone trying to help bumped into me, and I dropped them all again. Everyone turned to stare at me.

I placed the heap of mixed-up files and papers back on the table and looked around the room. All heads looked down at papers, carefully avoiding me. Now, all over Tanzania, budding businesses might remain nameless for months or years until this mess was sorted out. The economy could collapse. Hadn't I caused enough trouble? I hated to admit defeat but I just had to go.

Tears dribbled from my eyes, sweat from my armpits, and my small pool of courage ebbed away. I didn't want any more skirmishes with these occupants of the bureaucratic labyrinth. Midday, time to close shop. As I crept downstairs, I paused at the door of the secretaries' room. The bald man was coming upstairs. He didn't yet know I had trashed his office. Seeing me standing blinking back tears, he gently pushed me toward the secretaries inside the room.

In a low voice, he told me, "Leave your papers with them. They will process them when they get around to it. My office will send you a certificate with the new business name."

I felt joy and apprehension in equal amounts. Was this a real solution? I didn't wait to examine options or feelings. I squeezed his hands and whispered, "Thank you." My desire to flee was so great that I did as he advised and placed my papers neatly on Fat One's desk with a polite entreaty and expressions of appreciation for her help. What could I do but trust in whatever powers spread order through chaos and bureaucracy?

David and I left the coastal city and returned to our cooler and drier home in the near-desert of Mangola. We started to use our new business name (without the "and") right away. However, in an overstuffed part at the back of my brain, there was a file cabinet labeled Unfinished Business. Inside lurked a scuffed file folder, Officialdumb, and in that hid the untyped form for Kibuyu (and) Partners. The whole exercise could end in a "goych"—our word for an unrealized expectation, the let-down feeling of a failed promise.

Months later, and still no name change in our post box. I typed up a standard letter imploring the business office to send us the registered name and sent it every month. No response. I dreaded another visit to hot, sticky, ugly Dar in another attempt to wrest our new business name from the offices. I knew I couldn't return to try again. I'm not sure to this day if David was amused at all the effort that I put into changing our business name to one with the image of a gourd, a kibuyu. With his sense of humor, he would have told me, "You're out of your gourd to go through all the trouble."

A year after the visit to the dismal offices of Dar es Salaam, we found ourselves in Arusha town doing chores. We almost literally bumped into our seldom-seen attorney on the street. "Hello," said this urbane, elderly Indian gentleman.

"Well, hello, Mr. Kapoor, good to see you. We haven't seen you for months. How is the family?" Mr. Kapoor had helped us register our partnership "David

Bygott & Co." ten years earlier. Since then, we'd seen him only on rare social occasions. While we made polite conversation, he suddenly said, "Oh yes, you'd better check at my office. There is a letter for you. It's been lying around for rather a long time."

Oh dear, I thought immediately, what summons, or difficulty was ready to pounce on us? Taxes, fees, revocation of our business license? We grimly climbed the stairs to his office. And yes, the letter was from the Business Registration Office, certifying our duly registered name, Kibuyu & Partners. The man in the office in Dar es Salaam had kept his promise, and I would try even harder to keep mine.

Cyperus papyrus - head

CHAPTER 6
TEA WITH TOMIKAWA
A GENTLEMAN WITH A BROAD VIEW OF AFRICAN SOCIETIES AND AN INTIMATE VIEW OF MANGOLA: 1987–1997

"Take tea with me?" My question came out more like a plea than the cordial invitation I'd intended. I was practicing welcoming manners on my Japanese visitor, Professor Doctor Morimichi Tomikawa.

"Thank you very much," he said, "but not now. I must go back to Mama Rama's place before dark."

I tried once more; I really did want him to come for tea. "David and I would be honored if you would come for lunch tomorrow and have tea then." I made a little bow to emphasize the sincerity of my invitation.

Dr. Tomikawa (M.D. and Ph.D.) looked at me through his smeared spectacles and nodded. "Yes, tomorrow, I will come."

We admired and liked Morimichi Tomikawa. He was a long-term researcher who had come to do research in Mangola focusing on the Datoga tribe. Unfortunately, the majority of the papers he wrote were in Japanese and his results not widely known. I hoped a shared meal would be a chance to get to know him and glean insights about Mangola and the Datoga.

I decided to make the meal special, Japanese style. That meant enjoying the preparations, honoring the guest. This was part of my self-betterment process, trying to do things in a ceremonious manner—calmly, focused, with pleasure. My carefully designed menu included precious tinned sardines (a gift from Yoichi Wazaki, a colleague of Tomikawa's) with bite-sized pieces of fresh bread. I would prepare Mangola onions in a soup served with crisp carrot and cucumber slices from our vegetable garden. The main dish was chicken cooked in coconut milk, accompanied by rice balls wrapped in seaweed (another Wazaki gift). We also would have corn kernels stripped from the cob mixed with chunks of red peppers and zucchini squash, all cooked together with fresh ginger. Dessert? Flavorful, fresh papaya halves from my hybrid plants.

And of course, there would be Japanese green tea (yes, Wazaki again). I'd have the chance to put out my little Japanese teapot with the handle-less cups I'd received during a primate conference in Japan. I was inordinately proud of keeping the souvenir through several moves and continents. It reminded me of the courtesy inherent in Japanese culture, as well as my desire for ritual and ceremony. The meal for Tomikawa gave me a chance to add a touch of formality, something our rugged lives in the East African outback lacked.

Our guest duly appeared, nodding and bowing. We sat down to lunch at our favorite spot under the shady tamarind tree in front of the guesthouse. Tomikawa sat straight-backed with a gentlemanly manner. To me, he was the living definition of dignified and polite. His self-effacing smile and spectacles that needed cleaning made him seem vulnerable, too.

The soup and sardines slipped down quickly while we exchanged pleasantries. Over the chicken, rice, and vegetables, we listened to Tomikawa's history of his time as a researcher in Mangola.

"I came in the early 1960s," he explained, sipping his tea. "I stayed for three years, doing medical service as well as studying Datoga customs. They had their cows, and only a few people farmed. One subgroup of the Datoga tribe settled here in Mangola in the 1940s after the Maasai defeated them and took the best grazing grounds in the highlands. I came to study this local group."

"What did your research focus on?" David asked as I poured more tea.

"I wanted to know about the Datoga lifestyle and how the other local tribes related to one another. I am interested in multi-ethnic regional systems." He said it as though we understood that bit of anthropological jargon.

"Can you give us an example of," David paused to get it right, "multi-ethnic regional systems?"

"Mangola is a good example," he told us. "The relationship between the Iraqw people and the Datoga, for instance. They are very different in genetics, languages, cultures, and history, but for decades there has been intermarriage. How do their customs adapt? How do women fit into their husband's tribe? Who inherits the cattle? What system of justice do they follow? To find answers, I needed to understand the way these two tribes influence each other."

CHAPTER 6: TEA WITH TOMIKAWA

We let Tomikawa eat and drink while we thought about regional systems. We didn't want to overwhelm our guest with questions but were eager to learn more. In particular, we wanted to hear more about the history of Mangola. The Professor delicately nibbled a rice ball, swallowed it. While he wiped his spectacles with his napkin, David slipped in another question.

"What was Mangola like in the sixties?"

"When I first came to Mangola, only a few hundred people lived here: all spread around. That was before President Nyerere's efforts to make people move into villages. The tribes here in Mangola still mostly kept to their traditional lifestyles. The Hadza foraged, hunted, and gathered. The Iraqw had houses and farms here and there near springs. The Datoga moved their flocks and herds seasonally with a few compounds here."

I was glad to hear the names of three major ethnic groups mentioned. We knew individuals and families from all of them, most still living the way they had for hundreds of years. I was preparing to ask about the most numerous group, the Bantu, when Tomikawa mentioned them.

"While most of the people in Mangola belonged to those three different tribes, the Bantu came too. They settled here during colonial times."

"Yes, what about the colonials, the non-Africans. How did they fit in or not?" David asked.

"Colonial Germans came here on and off through both world wars. They didn't make much of an impact on Mangola. Officials came to hunt and left. European researchers like the Kohl-Larsens came and stayed for short spells. After World War II, the foreigners—mostly Germans and British—stayed up on the slopes of the mountains growing coffee. The colonialists hired people from different local peoples, especially the Iraqw. But most came from Bantu tribes."

Tomikawa paused to sip tea. I thought of Wazaki telling us how the Bantu tribes in Mangola thought of themselves as Waswahili, Wazaki's example of group identity. Bantu ethnic groups each had its own special language though they shared a basic structure. Language couldn't be the only factor that made them identify as a group. Bantu people were mostly farmers, but also have other ways of making a living. Did they have a shared history? I thought again of the American system of lumping diverse European tribes together and calling them White. That formed a caste system based on skin color, not lifestyle, shared culture, or language. I put the complex puzzle of group identity aside and listened to Tomikawa.

He put down his cup and took up his history lesson. "The Bantu came from all over the country, from Sukuma land, Kilimanjaro, Tabora, and Morogoro. The colonial masters hired farmworkers on their highland estates. Soon the ever-expanding Bantu families started coming down to Mangola from the highlands. They settled here to farm maize and grow onions. Mangola was the hinterland in the middle of the century. The Bantu tribes are here, they brought Swahili with them. They are mixing their cultures and traditions but only partly mixing their genes with the other ethnic groups."

Tomikawa paused before saying with a small smile, "Mangola is a pioneer place. This region is very multi-ethnic. And Gorofani village is an excellent example of the mixing of tribes."

Tomikawa wasn't the only researcher who saw Mangola as a special place for ethnic groups. Mangola attracted a variety of researchers. We had met many. In this cul-de-sac on the African continent, researchers could find four distinct groups—culturally, genetically, and linguistically different. These groups represented the four major African language groups: the click-speaking Hadza, Afro-Asiatic speaking Iraqw, Nilo-Hamitic speaking Datoga, and Niger-Congo speaking Bantu.

The tribes living in Mangola also still practiced a semblance of four different traditional lifestyles: Hadza foragers, Iraqw agro-pastoralists, Datoga livestock herders, and Bantu farmers.

Mangola was and still is a uniquely multi-ethnic mix. This mix attracted us as well as researchers. The history of these peoples and their ancestors also attracted archeologists and medical people. For our part, we mined all the researchers for nuggets of information.

David changed the subject from the complexity of tribal groups with a deft local question. "Our village. It has two names. We call it Gorofani. Maps call this place Ghang'ndend (ghung-n-dend)."

"Yes," said Tomikawa, still smiling. Qhang'ndenda is the Datoga name for the Chemchem Springs. "But Gorofani is so much easier to say!"

"It sure is," I said. "Is it true Gorofani is a corruption of the name of a man named Goldenfahn?"

"I'm not sure of his real name," Tomikawa told us. "He was possibly a German Jew or even a Pole. He came to Mangola after World War II in the 1950s and left before I came."

"He must have been a busy fellow," said David. "Goldenfahn left a lot of other traces as well as his name. We've found the vehicle tracks over the hills to Mbulu and Oldeani."

"And some of the fruit trees he brought here are still alive," I added.

"Yes," agreed Tomikawa. "The cement platform at Chemchem was where he had his sisal machine. He paid workers to collect wild sisal, shred it, and make rope from the fibers. You can also still see the citrus trees he planted at Chemchem and the big mango tree. He planted papayas, too, and sold the sap for the meat tenderizer, papain."

"The descendants of those papayas are still around," I said and bragged shamelessly. "The trees have adapted to Mangola's climate and soil. I've been planting their seeds and seeds from Hawaiian papayas. I'm crossbreeding them. I hope to produce an exceptional papaya, one with the hardiness of the large hardy local type with the sweet taste and fragrance of the softer smaller foreign papaya."

"And it's been a great success," said David. "We have the best papayas anywhere."

CHAPTER 6: TEA WITH TOMIKAWA

Papaya fruit

Over more tea and perfect papayas, we talked about the current status of Mangola. We shared a concern about the influx of people in general and specifically how the onion farmers took land needed by the traditional foragers and livestock keepers, who had fewer and fewer places to live. Tomikawa was concerned that many of the Datoga families he'd known had already settled down in the village.

"My old friend Gida Bashki has a large compound. It's surrounded by onion growers, houses, churches, the mosque, and the school. His sons take the livestock to distant places. Many of his family stay here in Gorofani year-round now. Some even grow maize and onions like the Bantu and Iraqw." He paused and shook his head. "Things have changed so much. Gida Bashki is old now. I am getting old, too."

Tomikawa did not strike us as old; he was energetic enough to make multiple journeys from Japan to wild Mangola and had the stamina to endure long field trips over terrible roads. He never told us exactly why he made the trips. We thought it was because he loved Africa and liked to visit old friends. Despite his senescent self-image, Tomikawa kept coming back. He made brief appearances, stayed for a few days with his old friend Mama Rama, then left to visit other friends.

He and Mama Rama had a special bond. She had housed and befriended him and his wife during all the years he'd been coming to Mangola. In addition to being a friendly base, Mama Rama's place was closer to Tomikawa's friend Gida Bashki. We knew Bashki as the esteemed Datoga elder in our village and Tomikawa often talked fondly of the old man.

Bashki died while Tomikawa was away in Japan. The next time the professor returned to Mangola he looked much older. I began to worry about how many more times he'd make the journey from Japan. After exchanging greetings, he said, "I have come to ask your help. I have a guest with me. I want to take him on safari. You have written the best guidebook for the Ngorongoro Conservation area. The book is an excellent source, but I want some special advice for this safari."

His praise flattered me so I tried my best to give him the pros and cons of camping versus lodges, where to go, and so forth. But as we talked, dusk and mosquitos descended on us and the professor looked ever more tired. I tried to maneuver Tomikawa indoors, saying, "Please, Professor, take tea with me. Come up to our house, away from these pesky insects."

Tomikawa looked down at the mosquito poking its syringe into his arm and calmly slapped it. He looked so distracted that I took his arm and led him into our sitting room. I lit lamps and set the kettle to boil. Tomikawa sat down slowly. He took off his dust-clouded spectacles when I handed him a clean cloth, a little ritual we'd developed. He glanced around our living room before launching into the subject on his heart.

"Gida Bashki is dead," He said flatly. "I had hoped to arrive before the death of my friend. I have come too late. Now there will be a big bung'eda ceremony to commemorate him." He sat quietly, polishing his glasses, staring off into the future. "I don't know if I'll be back in time for the important event, but I will try."

I prepared green tea and set out the pot, cups, and ginger cookies. Tomikawa leaned over the table, sniffed the steam, and smiled at some memory. He sat up, still looking distracted, and said, "I would like to see the ceremony but now I must return to Tokyo. I have been asked to teach at the university again. This work does not appeal to me. I am tired, but I cannot refuse. I may not be able to return to Mangola; my doctor warns me of my heart problems."

We returned to the subject of his guest. "I am here only for a short time. I will show my guest Mangola tomorrow; then, we go on safari."

Ah, so, I said to myself, another chance for tea with Tomikawa. "Please come with your guest and have lunch tomorrow."

"Yes, thank you, yes, we will come," and rose to go.

The two distinguished professors arrived on a breezy afternoon. I wasn't too surprised when I saw that Jumoda had invited himself along. Tomikawa had known Jumoda since he was a boy. The wise professor had recommended him as a potential guide and assistant to other researchers who came to Mangola. He'd told us, "Jumoda needs development." And indeed, Jumoda did develop. He seemed enthralled by Japanese researchers. He respected them more than any other tribe of researchers, whether Dutch, Irish, German, Scandinavian, Australian, American, or Tanzanian.

I brought my three visitors to the outdoor table under the tamarind tree by the guesthouse. The fragrance of nearby yellow barked acacias gave the setting an especially sweet and benign ambiance. While the men sat down, I prepared tea in my little Japanese teapot with matching cups and put out a bowl of sugar. Unlike with British tea, I didn't have to find lemon slices or milk because I had neither. I smiled at my guests as I poured tea into the cups. Jumoda tried to help by topping up the cups of tea with sugar from the bowl. Neither Tomikawa nor I wanted sugar and just barely deflected the heaped teaspoon; we preferred to have a beer and let our tea sit and grow cold.

I served my guests seaweed soup, followed by gingered beef, rice, and our garden vegetables. While we ate, we chatted politely in Swahili about the village, the government, and academics. Then I noticed something wrong, Jumoda sipped his tea frowning. He poured some more tea for Tomikawa's guest. The man was an eminent linguist who sat quietly watching the monkeys frolicking in the acacia tree nearby. He and Jumoda did not drink much tea.

Thinking I should give him fresh tea, I dumped out the cold brew and started a new pot. Placing it on the table, I turned to try to catch something Tomikawa was saying and, with an indelicate swing, knocked my beer off the table. As I leaned over to pick up the bottle, Tomikawa leaned over to help. Our heads hit one another, and in his haste to regain his composure, Tomikawa knocked his full plate of food off the table. We smiled politely at one another, very embarrassed. What could we do but laugh?

So much for tea ceremony, Mangola style. I cleaned up the mess, refilled glasses and plates, and we all carried on eating and talking. I saw Jumoda deftly put more sugar in his and our guest's teacup, but neither seemed to want to drink the tea. I told myself I'd better check the packet. But it wasn't the tea. Only later did I discover that the "sugar bowl" was full of sand, the same color as our local unrefined sugar. It was a bowl I used for incense sticks! They had been trying to sweeten their tea with sand.

After this amusing meal, I realized once again how socially vulnerable I felt in Mangola. I suffered from a dilemma—a dread of having guests combined with an eagerness to have them. I felt a childish letdown by my performance of Tea with Tomikawa. I could almost hear my father telling me as a young girl, "You take things too seriously, PeeWee. Relax, smile, no one is judging you. Just do the best you can." I had tried, and I decided that yes, I needed to smile at myself more.

Tomikawa did return again but not at the right time for his friend Bashki's ceremonial. On this return visit, we met him at Mama Rama's, looking pale, alert, and determined. He told us he had been in the hospital with the foretold heart problems.

"I think this will be my last trip to Africa," he told us. "But other researchers and students will be coming again to Mangola. They are keeping Japanese research alive here. I am keeping myself alive long enough to say goodbye to some very dear friends and places."

He asked us if he could use our car for a trip to the Lake Victoria region. We readily agreed. David had to do some work on the vehicle before it was roadworthy. When Tomikawa came to see the car preparations, I steered him away for some tea and goodies. Our teatime conversation was about writing, a subject important to us both. We discussed the difficulty of finding time to write and commiserated on how few people would read whatever we wrote. I told Tomikawa about my aim to write about our experiences in Mangola. He looked at me through those foggy glasses that I always wanted to clean for him.

"If you write about Mangola," he advised me, "portray it as a part of the whole of the Eyasi area. Tell about the different tribes living here and sharing this land. Tell about the Hadza and their ancestors, the foragers, hunters, and gatherers who lived here for thousands of years. Tell about the ancient rock paintings and irrigation canals, the bones, and tools of ancient peoples. And mention the modern tribes, Mangola's unique cultural mix.

"The history of this place is complicated, vibrant. People nowadays don't seem to appreciate complex, remote places like Mangola. Your stories can help readers understand cultures better."

He must have read the self-doubt in my face and added with a smile, "Try your best." He sounded just like my father. If you are reading this story, you can tell I tried to follow Tomikawa's advice. I hope our stories contribute to his "multi-ethnic regional" worldview.

After his trip, Tomikawa came to return the car and say goodbye. His body drooped. His glasses were so smeared that he let them slip down his nose. He looked over them as he raised his hand to focus on his cup of tea. I placed a handkerchief in front of him. He picked it up to wipe his spectacles, smiled, and said, "I have brought you a gift."

He handed me a beautifully wrapped package saying, "This is from my wife, Aiko." I was puzzled. Tomikawa had only mentioned Aiko briefly on rare occasions. I unwrapped the gift slowly. Nestled in paper was a lovely, enameled tray with an orchid design. I shook my head, bemused and pleased. "Thank you. Please thank your wife. I like this very much."

Tomikawa bowed his head and blinked several times. "There is something else," he said as he handed me a little package—a half packet of special green tea. His eyes twinkled as he said, "If we drink tea together, we don't need any sugar in it, do we?" I was delighted; we laughed.

"I am leaving now," he said. "Thank you for your car. Thank you for the tea, the food, and listening to me. Tell David goodbye for me; I am sorry he is not here to thank him for all his help, too." We both bowed. He stood up straighter, as though pulling himself together. I looked at this well-dressed gentleman wearing a pale tan vest over a light-colored plaid shirt with a lovely wine-colored cardigan. I wondered if his wife had made the sweater. A dark green hat matching his green trousers completed his natty outfit. I smiled at this white-haired man and wondered if his heart would be healthy enough to bring him back to Mangola again.

Tomikawa didn't return for Bashki's bung'eda. In 1997 he joined his old friend in death. It had been Bashki's wish that Tomikawa would be buried near him in

Mangola. Realizing that wish caused a conflict between Jumoda and other Japanese researchers plus aid workers and politicians too. Another multi-ethnic story for another time.

On a visit to Mangola in 2014, we went to see the burial spot for Tomikawa's bones. The spot was close to Jumoda's house. He guided us across a barren expanse to a stunted tree and pointed out the memorial plaque not far from the burial site of the elder Bashki. At least their bones could rest peacefully together. Meanwhile, the future would forge ahead and forget these two gentlemen of distinction. I wished I had some Japanese green tea to pour on the graves as a blessing.

Tomikawa's memorial in Mangola

A pair of Datoga gourds

CHAPTER 7
BUNGEDA BASHKI
A TRADITIONAL DATOGA DEATH CELEBRATION: 1997

Building the bung'eda tower

"Mzee Bashki amefariki!" exclaimed Jumoda, jumping off his bicycle as he swerved into our compound. *Mzee Bashki has died.* We stopped unloading our car and listened with full attention. Jumoda always caught our attention. His relative, Gida Bashki, was an esteemed elder, the local leader of the Datoga clan. He was also an influential figure in the Mangola area. Jumoda with all the vigor and self-importance of his youth told us the news in short Swahili sentences.

"Mzee Bashki died on Sunday. He died suddenly. He did not die in pain. It is a great loss. He was an important man. The elders sent runners to inform all the family. There are so many. They live everywhere, all around Lake Eyasi. Our clan is gathering from all over."

We stood, suitably impressed, and said, "Pole sana (po-lay sa-na)" That phrase is quintessential Swahili, compressing sympathy into two words. Everything from stubbing a toe to losing a favorite loved one can be addressed with the sympathetic words, pole sana.

Jumoda paused for a breath. I inserted in a question, "Does this mean there will be a bung'eda?"

"Oh yes, we're deciding when to hold the bung'eda," enthused Jumoda. "It's going to be a big event."

The traditional bung'eda is indeed an important event. We'd been to a few. A bung'eda is a Datoga ritual honoring and appeasing the dead. The word refers to both the burial ceremony and the tomb. This particular bung'eda would be elaborate because the man who had died was so widely respected as well as being the head of a large clan.

Like a news reporter, Jumoda continued to regale us with his news. "Gida Bashki has already been buried. Now the clan decides who does what. Who will build all the houses and enclosures for livestock? Who will cut the special posts to put on the grave? How much honey for beer? Where will we find the right cattle to slaughter?"

Gesuda gourds warming by the fire

Jumoda waved his arm at the distant mountains. He suddenly seemed to notice we stood sagging with bags over our shoulders, boxes at our feet, and more stuff in the car. I grabbed to chance to say, "Thanks for bringing the news. Please sit and relax while we finish unloading our gear."

Jumoda didn't reply but sat down on the bench; his eyes fixed on something distant. He was probably thinking about the work and cost of the elaborate funeral as well as the exciting events to come. Plans for the bung'eda would dominate his life and that of his clan for the next several months.

In the following days, we saw quite a few Datoga passing by our compound, walking on the road, and in the village. Apart from Jumoda, and some brief encounters with Bashki himself, I hardly knew any of the extended clan members. Even so, as friends of Jumoda's and also long-time resident foreigners, we had to pay our respects to the bereaved family. I decided I'd go to the clan compound with all three of our foster boys: Len, Sam, and Gillie. I'd also take Jumoda along.

Like all our trips, this one had to fulfill more than one purpose. Besides food gifts and tendering my condolences, I'd send Jumoda and the boys to the grain mill. I'd offer the use of our car to take the clan's grain to the mill as well as ours. Jumoda would get practice driving, and our boys would have a useful outing.

CHAPTER 7: BUNG'EDA BASHKI

Jumoda drove us over to where his clan had gathered. I was pleased he didn't drive into a ditch, or hit a tree, person, or dog. He didn't even honk the horn once. We piled out of the car. I looked around at the busy scene. People sat, walked, talked, and carried things from hut to hut. It looked festive, despite all the unsmiling faces. Datoga people always seemed to look rather grim, even without the double tattooed circles that many still wore around both eyes. They were serious people, with lots of social rules that I'd learned about from Professor Tomikawa. Even so, when around groups of Datoga, I felt as ignorant as a child at a cocktail party.

I greeted a group of women cooking huge pots full of maize porridge. I didn't know how many wives Mzee Gida Bashki once had, but at least three older women were in the compound. I scanned around and noted a dozen women and twice as many men. I saw no children except for a few girls helping inside the houses. I asked Jumoda to point out the female head of the clan, the first wife.

I set my gift of ten kilos of sugar down at the feet of the oldest woman. She nodded solemnly. Another older woman came close, weeping. Jumoda whispered to me, "She is the sister of Mzee Bashki." These women were expected to let their hair grow, a sign of mourning. They would lament vocally for days, even weeks.

Leaving the women, Jumoda took me to the men's area. This courtesy always embarrassed me, and I kept my head down respectfully. I felt awkward, a short foreign woman in the territory of strong local men. Jumoda introduced me to the eldest son of the deceased, his other sons, brothers, cousins, nephews, and more distant relatives and friends. I was so ill at ease, I made no effort to keep track of their faces, let alone their names.

Jumoda showed me the gravesite. I was surprised to see it was within the main compound. A pile of earth rose from the barren ground with some twigs and a large broken pot on top.

Jumoda pointed to the pot and told me, "That is a symbol of the man's life. It's on top of a mound of soil, connecting him with the earth. The men of our clan will slowly pile more soil, more twigs, branches, logs. They will place whole tree trunks around the mound over the next several weeks."

My eyes flicked from the mound to a man I recognized. Gwaruda, our night watchman, was there digging postholes for the new house of the eldest son. He'd already missed several days of work at our place because of helping during the bung'eda preparations. And these were just the early stages. I was sure that the dead man would have loved all the hustle and bustle.

I gave my respects and sympathy then offered to send my car with any grain to the mill. The men nodded at one another, pleased at the offer. Two of them rose and helped a woman haul a big leather bag of the kind that donkeys carry. It was full of maize, and both men strained to hoist it into the car with the help of Jumoda and our oldest boy, Len.

I was now free to head to a less demanding destination by myself. I took the cattle path leading towards the Chemchem Springs. The shade and screen of thick bush helped ease my tension on my way to see Mama Rama. Before crossing the main road, I stood looking over at her unfenced compound, so different from the one I'd just left. Children of all ages ran about playing and singing. Goats nibbed and chickens pecked whatever they could find. People milled around sitting here and there, chatting, eating, smoking, and drinking.

I surveyed the cheerful and confusing scene, smiled, and boldly joined in, finding Mama Rama in the largest of her many different huts. We sat in the cool of her "news center," as she called it. After the opening pleasantries, she said with a hoarse chuckle, "Since Mzee Bashki died, the Mang'ati have invaded the village!"

Mama Rama was born into the Iraqw tribe. Her highland tribe had a long and intimate relationship with the Datoga, or Mang'ati, as they were often called.

"Those people are too intense and formal," she told me. "They are obsessed with their cattle and rules. And now they are obsessed with this bung'eda. The news of the death has spread fast. All the relatives must come. They know exactly what they must do and will inherit."

Mama Rama laughed, adding, "Of course, it's the women who will inherit a whole lot of work." I could sympathize. The men were responsible for the funeral ceremony itself. However, the women in the clan would have many more chores than usual. They needed to gather supplies and prepare houses, rooms, beds, meals. "Ah, such a lot of work," sighed Mama Rama.

Jumoda kept us informed about the prolonged funeral arrangements. He also asked me to photograph parts of the ritual. That request surprised me as much as if he had asked me to braid his hair. I was an outsider and a woman. But I was also flattered, so agreed. David decided he would come along and take photos as well.

One particular duty for our watchman Gwaruda was to tend a patch of grass. On the day of Bashki's death, the men chose a small area of grass on the stream bank. That turf would play a part in the final ceremony. A man of good heart, respected by all, was appointed to protect the grass from animals or people. Gwaruda was honored by the position.

We were rather proud Gwaruda was chosen, yet it meant that he'd be dividing his time between his paid job with us and his unpaid bung'eda jobs. He'd be busy, with little time for his own family. I'd have to help them out with some supplies and cash. The thought showed me how the chores spread out from home base into the surrounding community.

Among the items required for the funeral were a large number of fresh posts, cut from living trees of seven different kinds. All but one had to be a species of the hardy commiphora. These twisty, thorny trees exuded fragrant resin related to the famous myrrh.

Jumoda told us that the next stage was the slaughtering ceremony. "An ox will be killed. People will eat the meat but keep the fat. That will be used to anoint Gida Bashki's sons who will climb the mound. We are building the tower now," said Jumoda. "Come over tomorrow and take pictures."

I went as requested and photographed men carrying tree trunks to contribute to the structure. Over the conical grave, they built a small, tapered, upright cylinder of logs, plastering them together with mud mixed in a pit nearby. Over time the

mound would resemble a tomb we'd first seen after our arrival in Mangola. We'd come into a clearing and stood still, puzzled to see a mound like a big termite hill, ribbed with tree trunks sprouting leaves. It resembled a huge conical mud cake with leafy green birthday candles stuck in it.

I left the building site before evening as visitors and family started singing, dancing, and drinking. They repeated this for two or three days. Then all the visitors headed home until the next event. Similar meetings took place over the next couple of months. Each time, the men brought more logs, building the tower higher and broader.

One day, Jumoda came to ask if I would take some tins of honey over to the clan compound. Honey was an essential part of the ceremony, needed for making plenty of honey beer, called *gesuda*, a delicious and alcoholic concoction. I agreed at once; I'd always wanted to see the honey beer operation.

At the side of the compound stood a large, open-sided, airy hut—the brewery. A half dozen big, shiny gourds stood around a small fire. I learned that the brewing process called for two 20-gallon tins full of honey poured into several large calabashes. The brewers added water and a particular root that accelerated the brewing process. The liquid fermented as the gourds got turned and rotated around the fire by men responsible for guarding the liquid treasure.

The final days of the ceremony took place some five months after the death. David and I joined the mob who'd gathered from afar. Despite the black cloaks on the men, it was a colorful occasion. The women wore new leather skirts and masses of beaded necklaces. Their brass bracelets on arms and legs gleamed. The clansmen brought the final round of logs—smooth commiphora trunks in pastel shades of grey, green, and blue. Many golden-brown gourds full of gesuda mead warmed their round bodies near smoky fires.

Several youths brought the black sacrificial bull, which had been kept and fattened at a compound some distance away. They slaughtered the bull a few yards from the bung'eda site. Then the women took over, skinning the animal and cutting it into chunks.

Women butcher the black bull

Meanwhile, the dancing had begun. Young men and women spilled outside the main compound to form groups. They made quite a din, emitting cries, their feet pounding the dusty ground.

There were two main kinds of dances. Most prominent and popular was a jumping dance composed of a line or semicircle of men who stood opposite a lineup of girls. Each man wore his black cloak, ornamented with white buttons and metal decorations. Various ornaments signaled the social status of the men. Reputations of courage and success counted, especially with the young girls.

The unmarried girls were mostly in their teens. They looked like freshly baked cookies in their stiff leather skirts and cloaks. The older girls tried their best to act uninterested; they hardly ever smiled. Professor Tomikawa had told me that Datoga women had to defer to and show respect to men at all times. They must not

CHAPTER 7: BUNG'EDA BASHKI

Datoga girls dressed for the ceremony

look directly at men. When married, they had to keep their gaze away from their fathers-in-law and other male relations. Smiles were not allowed.

The young ladies in the bung'eda dancing group certainly did not stare at the men. Indeed, they seemed more concerned about their dresses than about the warriors. They clustered about, checking each other's finery. Most wore handmade leather outfits—skirts pleated or gored to fit over the hips, with a gathering above one hip, a cape hanging off one shoulder with lots of beads accentuating the asymmetry. Very stylish, I thought.

When the girls jumped in the dances, they could hardly rise from the ground, weighed down by their adornments of brass and beads. With difficulty, they tore themselves from earth's gravity, not radiating happiness but looking determined and solemn.

A dance master beat time on a leather shield with a stick. He chose which men could dance. From time to time, some brave warriors ran toward the girls with whoops and shouts. When they rushed forward, a few would lay their sticks on the ground. The dance gave the men and their future brides a chance to assess and admire one another.

The other type of Datoga dance was very different, full of humming and percussive rhythms. A visiting group performed about fifty yards from the jump dancers. The men clapped and made ululating, rhythmic cries.

After watching the dancing for a time, we looked for shade. We rested under an acacia tree to drink and eat. The dancers kept on. They'd dance into the night; the thought made me tired.

We returned to the main compound in the glare and heat of late afternoon. A new group had come to finish the building of the tower. The edifice had become an imposing pillar, as high as a tall man with a child standing on his shoulders. This imaginary child could peer into the center of the tower. He would see a cavity full of soil, the dead man's symbolic head.

At sundown, men brought large, forked branches and leaned them against the tower to act as ladders. Young men brought dry cow manure from a nearby pen, carrying it in stiff goatskins. They handed it up to men standing on top. The "kings of the mountain" poured the manure into the cavity and tromped it down.

A crowd gathered around the structure and black-cloaked men formed a circle around its base. They held their sticks horizontally at waist height to create a barrier. If anyone strayed into the arena or pushed too hard, the guards took a stick and drove them back. At this stage, I saw no women. They had all gone into the houses, not permitted to watch the ceremony.

Warriors hoisting grass up to the bung'eda tower

The procession began. First came the grass-bearers—four men carrying a litter containing the chosen turf. The men who dug it up used wooden digging sticks; no iron was allowed for that part of the ceremony. I noticed that the grass had been covered with earth to keep it cool and fresh.

With great care, the men handed the clump to others on top of the mound. Those fellows spent a long time combing out the dead man's hair, the long strands of grass. They arranged them to trail over the sides of the tower.

Next came a comic interlude with two older men. We recognized one of them, Gidesh, a hillside neighbor. They hauled a long, thick, gnarled vine supposedly rep-

CHAPTER 7: BUNG'EDA BASHKI

resenting the dead man's belt. They struggled to wind the vine around the middle of the tower, stumbling and tottering. We didn't know if they were drunk or just acting. Maybe both. Most of the men had been drinking deeply of the delicious honey mead. It took some time and a few helpers to wrap the vine into a belt shape on the tower. The audience laughed softly each time the men stumbled.

The finale came as the sun neared the horizon and the air started cooling off. A procession of men and boys entered the circle: the first sons born to each of Bashki's wives. Each male wore only a leather kilt around his waist made of new hide prepared by a close female relative. The pale new leather contrasted with the old worn brown skirts the other participants wore.

The sons walk to the tower, carrying saplings.

Sons planting saplings on bung'eda tower

Each man or boy carried a slender, leafy sapling on his shoulder. All the boys, youths, and men took off their rubber tire sandals before climbing the tower. At the crown of the tower, each son planted his sapling, then descended on the opposite side.

One boy, in confusion, started down the wrong side but was quickly corrected

Men clashing their sticks together

by the watching elders. When the last son had descended, the orderly circle of crowd controllers broke up into a milling mob of men with sticks raised above their heads. Singing a farewell song, they surged around the tower once, clacking their sticks in the air. They all charged out of the boma, clashing their sticks. Soon they rushed back in again and beat their sticks together as they circled the tower. Suddenly they stopped, and the circle broke up. Spectators started to leave.

The memorial tower, dressed in the elder's memory, looked serene but pleasing in the twilight. You might have heard Bashki chuckle or grunt as he settled down for a long rest. We thought this was the final part of the ceremony and started to leave with the others. We stopped when we noticed the women appearing. They emerged from their houses in a somber group. Most wore dark cloaks and had

Women pray for the dead chief

shaved heads; others wore new pale leather skirts. They stood beneath the tower and started a beautiful, melancholy chant. Jumoda told us that the women were singing a farewell prayer to the deceased spirit.

In time, the spirit tower sprouted leaves, and the family compound moved to a new spot. We passed by the bung'eda memorial sometimes. When I passed alone, I often stood near the tower to listen. I'd look at its leafy arms waving at the blue sky with its roots deep in the earth. Was there a spirit resting inside? When the everlasting Mangola wind swished the branches, the mound seemed to murmur. I could imagine a low voice telling me secrets about life and death.

The elaborate way the Datoga dedicated their dead to the earth struck me as a soothing and satisfying ecological burial. The long-lasting bung'eda tower became a symbol of life's transience and renewal.

The bung'eda tower, some months after the ceremony

CHAPTER 8
GUDO THE GUIDE
AN UNUSUAL HADZA MAN: 1986 ONWARDS

Gudo

Gudo, our Hadza friend, had a sense of style. When he first arrived on the steps of our stone house at Mangola Plantation, he looked amazing. He wore a red and yellow flowered shirt, with a natty charcoal waistcoat and grey and black striped trousers A jaunty blue cap topped his curly-haired head. On other occasions, he wore a zebra mane hat or a headband with feathers in it. Often, he'd show up in baggy shorts, an old t-shirt, and flip-flops.

But Gudo's appearance did not reflect the man we came to know. He wasn't nearly as flamboyant as his attire. Gudo was rather quiet, serious-minded, and resolute. We got to know him after he came to see us about getting work. At the time, we lived on Mangola Plantation a little distance from the sprawl of Gorofani village. David was again away on a safari, so I welcomed Gudo myself. I knew he was a member of the Hadza tribe. While I had a keen interest in the various tribes in the Lake Eyasi area, I found the Hadza tribe one of the most interesting.

Gudo sat down and I brought out my usual offering of peanut butter and jam sandwiches with tea. We began a conversation in Swahili.

"What kind of work are you looking for?" I asked him.

He answered in a voice raspy with the smoke from ten thousand campfires: "Research assistant."

Gudo didn't look much like a research assistant in his colorful ragged clothes and flip-flops. But I could imagine him in that role more easily than cocky, happy-go-lucky Kaunda, the resident Hadza who helped the around the farm. As on cue, Kaunda appeared around the edge of the porch. He and Gudo clicked their hellos and exchanged news. All the Hadza knew one another well—there were fewer than 1,000 left, scattered over the Eyasi Basin and surrounds. I watched the two men talking and gesticulating. As usual, their buoyancy, good humor, and clicking language impressed me.

After Kaunda bounced away, I asked Gudo, "What sort of research do you want to do?" "Hadzabe," he said, using the plural word for the Hadza people.

"Have you worked with researchers before?"

"Yes. A long time, Ras, then Ana."

Those names didn't register at first. I looked at Gudo more carefully. Somehow, he seemed familiar.

"Have we seen each other before?" I asked.

"Yes," he said. "You came to the *Bonde la Makaa* (the valley of the charcoal). I was there with Peta and Ana." The names resolved themselves into ones I knew. Peta was Peter, and Ana was Annie. I tried to recall my little foot safari to their camp. They were researchers who had told us about the Hadza for years. Peter Jones was an expert on stone tools; we first met him at Mary Leakey's camp at Olduvai Gorge. Peter's wife, Annie Vincent, was studying the Hadza diet for her Ph.D. in anthropology.

I knew that Peter and Annie had a camp in the valley east of our village. I'd gone there to see her in action. A cheerful Hadza woman had guided me on a long walk over windswept and dusty slopes, through rocky thorny gullies to their research camp in the charcoal valley. I found Peter there, and he took me to the cliff where several Hadza women were digging tubers. I picked Annie out at once, a striking pale blonde among the dark Hadza. She was counting and weighing the large potato-like roots they'd dug up.

I looked at the heap and blurted out, "So many in this one spot! Are they all from the same plant?"

"Yep, one plant. And those aren't all the tubers," said Annie. "The women usually leave some tubers untouched and re-bury the exposed roots of the parent plant. That way, when they come back in a year or two, they can harvest more. These tubers are the mainstays of their food supply."

The foraging team returned to camp. Peter poured us some precious drinking water as we sat down on the teetering camp chairs or logs to rest. A strong acrid smell of burnt feathers permeated the air. I looked around and noticed that some Hadza men and boys were grilling small lumps of something on sticks over a fire.

"They're toasting the mousebirds they caught," Annie told me. "The birds have

Annie watches Abeya digging for tubers

been stuffing themselves with the fruits of the *madabi* bushes that grow in thickets all around. The Hadza shoot them and roast them, berries and all!"

I thought how delicious a squab-sized bird stuffed with orange, persimmon-like fruit would be. "Yum!"

"Yeah, these people do have a wonderful diet. They eat wild fruits, baobab seeds, fresh meat, tubers, honey, all sorts of things," Annie explained.

"Do you think the Hadza are healthier than other tribes?" I asked.

"Absolutely! They don't eat much maize meal or refined foods. Loads of studies have found that their health is exceptional—unless officials force them to live in villages." I almost missed that last comment because one fellow had caught my eye. He was creeping over to the bushes with his bow and a small arrow—just a long stick whittled to a tapered point and fletched with guinea fowl feathers. His first shot impaled a mousebird. That man was Gudo.

Now he sat on our porch. "What happened to your work with Peta and Ana?" I asked.

"They went to Arusha. I do not know if Ana comes back. I hear other researchers are coming soon. They come to study the Hadza."

"What researchers are coming to Mangola?" I asked, feeling ignorant. I thought I was better informed. Here was a man who lived in the bush. Where did he get the news?

"Ras and Niklas," said Gudo. "Ras," I immediately figured out was Lars, and "Niklas" was probably Nicholas somebody.

Lars Smith had introduced us to some Hadza people during a brief visit to his camp in the Yaeda Valley in the 1970s. Back then we were living in Serengeti National Park, studying lions. Our friend Hendrik Hoeck, who studied hyraxes, enticed us to go. After a full day's drive, we reached Lar's remote camp and stayed overnight.

The next day a couple of Hadza working with Lars guided us over the wild hills between Yaeda and the Eyasi basin. We crammed ourselves into one Land Rover and wandered through the bushland, through Yaeda Hills and Valley, reaching springs near the shore of dry Lake Eyasi. Our Hadza companions told us they hunted there because wildlife came to drink. I'd remember the springs 20 years later when visiting another researcher, Daniela Sieff, who studied the pastoralists who destroyed those springs. (see Chapter 16)

That first visit to the Hadza in the 1970s hooked us to the tribe. On that trip we left the lakeside and drove over the rugged hills in the golden glow of twilight. We made camp under a giant baobab tree where Hadza huts grew like thatched mushrooms in the idyllic setting.

We foreigners set up our tents and started our small campfire. I was surprised, puzzled, and pleased when some Hadza men brought me meat they'd scavenged from an antelope killed by lions. I remembered the event vividly because I thought the meat would go to the men first. Instead, the Hadza men gave it to me to distribute. That became my first lesson in Hadza sharing. I learned from Lars that older women often doled out the food.

That same evening the Hadza sang and told stories that Lars tried to record. The Hadza seemed a cheerful, happy group; they enchanted us. The short adventure made a deep impression on me.

Lars explained his research on the way back. His work focused on Hadza hunting and food sharing. To help him in his work, he'd hired a research assistant who could read, write, and speak the Hadza language. Gudo was the man. He became the primary Hadza research assistant.

When Dr. Mary Leakey was excavating fossil tracks at Laetoli on the southern plains in the 1970s, she needed a skilled tracker. Lars sent Gudo there to identify some of the tracks. He worked alongside Peter Jones. Peter introduced Gudo to Annie. And thus, Gudo passed from researcher to researcher.

Now, a decade later, Lars was returning, bringing yet another colleague, "Niklas." Certainly, Gudo would be the appropriate assistant. I told Gudo, "I will gladly recommend you. I hope Lars and Niklas come here, and I know where to find you."

Lars and Nicholas Blurton-Jones arrived the following week. Nick was based at the University of California at Los Angeles. He came to Mangola to start a study collecting basic information on Hadza numbers, births, deaths, health. That, we knew, would be quite a job, what with the changeable Hadza names, multiple marriages, and unknown birthdates.

Nick's project was ambiguously and carefully entitled "Human Ecology in the Arid Steppe of Arusha Region." The project provided an umbrella for different researchers' interests, as well as studies of other tribes.

I reckoned that Gudo would fit well under that umbrella. And he did. Gudo became a much-needed assistant for several different scientists.

We always welcomed researchers of any tribe coming to study local tribes. During our two decades living in Mangola, we met and hosted anthropologists, archaeologists, botanists, geneticists, geologists, linguists, and floundering students. In addition to the academics, many journalists, filmmakers, photographers, and casual observers came from all over the world. Mangola was a hot spot for interesting people in a fabulous landscape with a long history.

Gudo accompanied many of these researchers and helped translate from the Hadza language to English and Swahili. During this busy time of researchers coming and going, I began working on a project with him.

"I want to write down Hadza stories," he told me. "Ever since primary school, I've wanted to write Hadza stories and songs. Niklas told me to start doing that, but how?"

Here was a chance to help Gudo with encouragement and equipment. I gave him our little tape recorder, pencils, pens, papers, notebooks. I gave him a push, too.

"Start now," I told him. "Start recording stories. Use this tape recorder. When you can, listen again and write the stories down."

He came back after a week and said, "I recorded many stories. The batteries are dead." David put in new batteries and turned on the recorder to listen. Slurred words, groans, and grumbles came faintly from the tape. We couldn't hear the Hadza words; the wind snatched at the voices, making them practically unintelligible.

"Gudo, have you written down these stories?" I asked. He shook his head. "Hard to write in the dark, and the voices on the tape not clear." After several such delays and disappointments, we agreed that Gudo should write out the stories after he heard them, not try to record them.

Gudo came to our compound one day and handed me his notebook. "Mama Simba, here is my first story."

I was delighted for a moment or two. Then I realized I had no idea how to

decipher the Hadza rendering. At that time, Gudo could not write down the many click types and the spellings that he would learn later from linguist Bonny Sands. Luckily, Gudo had also made a Swahili version, but alas, even that version looked like chicken scratchings. Translating from Swahili into English would be hard work, and already much had been lost from the original Hadza story.

Ah-ha, I thought, I will teach Gudo to type. We had a spare computer I used to teach English to our schoolgirls. So, I set Gudo up in the smallest guestroom on our compound at Mikwajuni. There he could type out his stories in peace.

He learned more rapidly than I'd expected. Alas, Gudo had no concept of punctuation. His wordswouldrun onandonwithno periodsnocommas norany indicationofwheretostop or startitwas veryfrustratingwhen Itriedtountangle the-sentences. Gudo and I sat for hours trying our best to translate traditional Hadza stories first into Swahili, then again into English. I despaired of ever producing a book with him.

While working on stories with Gudo, I learned more about his early life. Gudo laughed when he told me how frightened he felt when forced into primary school in the Yaeda Valley. The government of Tanzania hauled youngsters away from their camps to go to a mission school run by the Assemblies of God. "That was in 1960s, I think," he told me, adding with a frown, "I remember feeling small when I looked up at the stern-faced Iraqw teachers.

"They taught Swahili, arithmetic, and English." Gudo chuckled as he recited the names of his fellow students, eight boys, and two girls. I was interested to hear that two Hadza brothers, Naftali and Richard, were part of this first cohort. They'd both become significant representatives of the Hadza tribe as adults.

I got the impression that Gudo and the others didn't learn a lot. They had to walk to school and back to their camps every day. There were no boarding schools. They had no money, no place to stay, and no transport. Getting to school became such a hassle that the boys and girls ran away from school for months at a time.

In the early 1970s, the local authorities decided not enough water existed in the Yaeda Valley for a school, so the Tanzanian government built a boarding school at Endamagha in the Mangola area. This tiny village was at least a long day's rough drive from Yaeda. During school terms, local authorities rounded up the Hadza children from their camps in Yaeda Valley and hauled them in trucks to the school.

Gudo was among those who went to Endamagha School from 1975 to 1976 along with Naftali and Richard. After primary school, this first cohort had to take exams and, as Gudo said, "We did OK." Then the boys had to go to trade school. Gudo told me, with sarcasm in his voice, "They asked me two questions during the interview: 'What is Swahili for someone without teeth?' and 'What is a sign of rain?' What stupid questions!"

We laughed and laughed at that. I urged Gudo to tell me what happened after he'd correctly answered such stupid questions. He sighed, saying, "Then they sent me to Bashay School. That was up in the Mbulu hills, really far from my camp in the Yaeda Valley. I had to walk up the escarpment before dawn, study at school, stay overnight, study the next day until afternoon and walk back to camp. I was always tired! After two weeks, I quit and went home."

"Was that the end of your education?" I asked.

"Well, after a month, the Mbulu beekeeper caught me. I went to seminars on bees for two weeks, then spent a month learning about ironwork. I returned to our camp, but what could I do? Where would I get iron? How could I get hives? I knew about the bee sicknesses. I knew that honey badgers rip open the hives, and people rob the hives, too! I could not keep bees. So, I just went back to the bush."

"And then you lived around Yaeda till you met Lars?"

"Oh, no! There was the army! When I was in school, they put my name in a register with the names of the other students. The army read the book. They found our names. They came to grab us to be soldiers! They took us to Monduli army camp. We arrived in the dark. Heh! The gate guards ordered us to get out and find bedding, beds, and uniforms. No one gave us any food. I felt trapped. They treated us like dogs. The officers shouted at us. 'Do this! Do that!' They blew whistles at us all the time."

"Did it get better?"

"No! It was terrible. We had to escape. A friend and I climbed over the fence at night. We ran west. We crossed the plains on foot during the night and hid out in the bush during the daylight. Up, up, up over the rift wall until we finally reached Mbulu plateau. We had tossed away our shoes and uniforms, so went barefoot, wearing only shorts. We ran back into Yaeda valley to eat wild fruits. We were so happy. Freedom!"

I smiled, shaking my head at the obvious glee Gudo exhibited, waving his arms around, crying out "Uhuru" in Swahili and "Freedom" in English.

"What about the others?" I asked.

Gudo shook his head sadly. "Those cowards, they were stuck in the army for two years! I was free hunting in the bush. Then one day, Ras came, and I worked with him. When he left, I went back to hunting. I married Elizabeti; we have the children. Then other researchers came. Then I came to you and got the job with Niklas." Gudo gave me a nod of confirmation. I nodded back. He continued.

"We have done things together. You and I started the bee project, too. Now this book of stories." He looked at me with his head tilted, as though assessing whether I was a friend, tutor, or researcher. I hoped I might be all of that.

By then, Gudo neared the age of forty. He looked healthy and proud, finally achieving his life's dream of making a book of Hadza traditional stories and songs. One of his stories, "The Hunter's Gift," my favorite, is at the end of this chapter.

In the introduction to his book Gudo wrote, "I went to Mama Simba to ask for pen and paper to begin to write our stories to leave our children with something of their heritage." Although he came to me early, his project only became realized because of the focused team effort. Bonny Sands had the linguistic skills to transcribe the Hadza stories and translate them into English. Nick Blurton-Jones got the booklet printed in the US. Piles of books came to Mangola.

The book contained nine tales, tongue twisters, and a section on Hadza history and lifestyle. Gudo illustrated the book with his colored pictures. He sold his copies to tourists; we sold copies for him at lodge gift shops. He was genuinely pleased, not only with accomplishing a project he'd wanted to do for so long, but with the money as well.

Drawing by Gudo: Seeta, the Moon and Haine, the Sun with their children, the stars.

CHAPTER 8: GUDO THE GUIDE

We never employed Gudo. He went with us as a guide, companion, and translator on many of our trips. We supported his efforts to help his tribe win respect and land. When he came to our house the first time, we met Gudo Mahiya. By the time Gudo put his name on *Hadza Stories and Songs*, he changed his name to G.G. Bala. But in the book, he states, "My real name, as I was born, is Mkune." We admire and love him, our intrepid friend Mkune Gudo Mahiya G.G. Bala.

Here is an abbreviated version of a Hadza story told by Gudo. You need to listen with full imagination as Gudo tells this tale in his rough voice full of clicks and tsks with abundant gestures and imitative sounds.

The Hunter's Gift

It was a big camp. An old hunter named Iyeye lived there; he was a real meat expert, but these days he wasn't getting any. He went to the place where the animals came to drink. He built a screen of brush to hide and settled down to wait. He tried to stay awake, but many times he fell asleep. He slept deeply, snoring away. In the morning, he found the tracks of eland, buffalo, zebras, and other animals. He was amazed that he had slept through their visits. He said, "Where did all these animals go? I will have to wait again tonight."

Iyeye went back to camp empty-handed. He returned to the hide the second night. But again, he fell asleep. In the morning, he saw the tracks, big eland tracks. He was surprised and wondered, why am I always sleeping?

His wife and children in camp scolded him, "Why do you bring no meat back from your long nights outdoors? It's because you're just lazy. You sleep all the time!"

He told his wife, "Give me roots, woman so that I can eat my fill. I am too hungry!"

So, she fed him till he was full because she was his wife.

"Now, give me water!" He drank it, then went back to the water hole. But he slept and snored while the animals brushed past the hide, chewed on the brush, drank, and left at dawn.

When he was making a fire, he saw a butterfly.

He called to it, "Come here, butterfly, I need a favor."

The butterfly asked him, "What do you need?"

Iyeye said, "I need to kill animals for meat. But I'm always asleep when the animals come. I need something to make me stay awake."

The butterfly went away and brought back what he needed.

That night, the hunter returned to his hide by the waterhole. He felt sleepy. Then he heard something at his ears, *buzzz, buzzzzz, zzzzz*. It bothered him a lot. He waved his arms. The night was long. He stayed wide awake, constantly slapping. His body burned from pricks.

Then came a big zebra. The hunter took careful aim and shot it. *Bzzzz, zzzzz*, went the insects round his ears. The buzzing and stinging upset him. The insects tormented him. An eland came; Iyeye hit it. More animals came. He shot and shot his arrows, and in between, when it was quiet, the insects ate him.

He cried out to the butterfly, "What is this thing bothering me? I haven't slept, what kind of thing is eating me?

She answered, "It is called a mosquito."

In the morning, Iyeye went to pick up the arrow shafts. He found a dead zebra, an eland, a buffalo, and further away a dead impala. He cut the skins and roasted the meat. He went back to camp to get help to carry back all the meat. They all asked him how he did it. He told them the butterfly had given him a gift to keep him awake.

The people sang his praises as they carried the meat over their shoulders. But when they were eating meat that night, many insects came buzzing and annoying them. The people started slapping, and the children cried loudly. Everyone had to move; they left Iyeye behind. But the mosquitos followed, and until today the mosquito eats us.

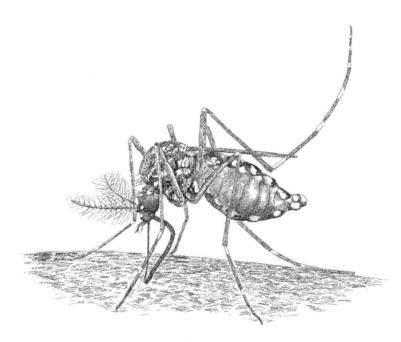

Mosquito

CHAPTER 9
TREASURE HUNTING
THREE ANECDOTES ABOUT MANGOLA'S ANCIENT ARCHAEOLOGICAL SITES: 1992

Miners looking for "Geman treasure"

We received an odd letter, translated here from the Swahili:

From: Mzee E.W. Mataro
Mangola 03 February 1992
To: Mama Simba, Mangola, Gorofani
REQUESTING TO KNOW IF YOU ARE ABLE TO WORK AT REVEALING GERMAN SITES WHERE TREASURE WAS BURIED LONG AGO.

If indeed you are interested, let me discuss this with you in private. I know of many places having mysterious kinds of things like cement and paint and German writing. These places are here in this area of Mangola and outside of Mangola.

I will be grateful for your reply.

I am Mzee Edward Warioba

I was away on safari at the time, so David decided to meet Edward and find out more. He sent Edward a reply and invited him to come to our compound. A polite man wearing clothes much too big for him came in the afternoon. He introduced himself as a prospector who had found exciting and secret things.

He told David in reasonably good English, "We don't spread the news about these special places, of course. We just pretend we are looking for gemstones."

"Well, what have you found?" asked David.

"There is a site near here. We have dug long and hard and have reached a layer of cement. We tried to blast through it with dynamite, but it resisted our efforts. There is writing nearby, and we need someone to explain it to us. We have heard that Mama Simba knows German. We think writing is German. Do you know German?"

"I know enough to know if the writing is in that language," David said. Edward leaned forward and whispered, "If we figure how to get through the cement, we may find gold, silver, or diamonds, even guns, maybe."

David had an inkling of what he'd find. We knew that many local people believed Germans buried their wealth before leaving Tanzania during both the world wars. The prospectors' faith in this myth withstood the total lack of evidence for it.

David picked Edward up on the edge of the village the next day. They drove up Baobab Hill and walked to a prominent rock outcrop. David took one look at the overhang and recognized it as an ancient rock shelter with paintings. His conclusion was confirmed when he saw circles and sunbursts painted in ochre on the wall.

In response to Edward's wide-eyed question, "What do you think they mean?" David told him, "Sorry, these designs are not words. Germans did not make them; no modern people made them. Ancient hands, maybe the ancestors of the Hadza or Sandawe people, painted them. This is modern treasure. These signs are *mambo ya kale*, the remains of people who stayed here long ago."

Edward tried to absorb this discouraging verdict. He then showed David another site with a pit about ten feet wide and six feet deep. He pointed to the floor of the hole.

"We know there is something under there. The medicine man told us we would find twenty-one boxes under here."

"How does the medicine man know that?" asked David.

"From the internet."

"The internet?"

"You know, the internet of all the medicine men. They know these things. Here's a piece of the cement, have a look."

"Sorry," David said again, "This chunk isn't cement; it's hard granite. It's part of this rock outcrop. There are rock paintings on the sites all over Africa. People long ago found shelter in these overhangs and caves. They drew on the rocks. There is no treasure buried in them."

Edward leaned against the wall of rock, his whole body sagging with this dis-

CHAPTER 9: TREASURE HUNTING

appointing interpretation of his treasure site. David went on talking to the numb man, "No one would bring their valuables out here to bury in the bush. The designs and stone tools are the wealth here. They are the remains of the people who lived long ago. Please, Edward, leave these sites to people who know how to excavate properly. Don't destroy these old treasures."

Edward may or may not have understood or accepted David's advice, because he didn't return after being dropped off at the edge of the village. However, the little adventure of the local treasure hunter reminded us of the vulnerability of the rock sites.

Living in a land of such archeological treasures, we wanted to see them for ourselves. Mangola was part of a vast and valuable zone of ancient sites in East Africa. We'd first seen singing stones, or gong rocks, when we lived in Serengeti National Park. We made the discovery for ourselves by hunting for tiny frogs that lived in the cups hollowed in the boulders. What a delight—

Rocky wonders, bang out a tune.
Pound out melodies, dance to the moon!

Later we'd seen rock shelters with signs and symbols at places throughout the Lake Victoria region. Those carved or painted rocks appealed to our visual natures. We liked the designs a lot more than the stones and bones requiring the mental energy and training of real archaeologists.

When we could, we went exploring for the rock art around Mangola. The safaris allowed us to be out and about in the bush, something we loved. Many of the rock shelters still had pictures of animals—giraffes, zebras, eland, rhino, buffalo,

Rock painting of giraffes, kudu and humans at a site in Yaeda valley

kudu. Sadly, many such expeditions led us to sites defaced and dug up. We often felt as distressed and disappointed as Edward had been about his lost treasure.

Trained archaeologists were few, rare jewels among uncut stone heaps of treasure hunters. One expert was Audax Mabulla, a Tanzanian doing his doctoral dissertation on local rock art. We got to know him when he brought his students on field trips to Mangola. Another visitor who taught us about ancient sites was Mary Leakey, a professional archaeologist. Mary worked at sites throughout East Africa with her husband, Louis Leakey. After Louis died, Mary kept working, mostly at Olduvai. We came to know her well during our years in Serengeti when we were studying lion ecology. We visited her at her camps at Olduvai and also Laetoli when she discovered ancient footprints made by human-like walkers over three and a half million years ago. She kindly hosted us sometimes at her home in Nairobi, too.

Mary Leakey in the field - windblown and intent on discovery.

After moving to Mangola, we saw her less often. She flattered us when she said she wanted to make the long dusty journey to Mangola to visit. If she came, we'd have a chance to glean insights into local archaeological sites. We felt thrilled and a bit intimidated, too. We worried about how we could entertain this famous woman. She'd published books, given lectures, trained students, and established a family based in Kenya. What could we possibly offer?

"An escape," Mary told me. "I have to go on safari with donors and guests to see

CHAPTER 9: TREASURE HUNTING

the Kolo rock paintings. I'll need an escape at the end. I can only stand so much of—well—work." Mary was nearly 80, a slight but hardy woman shriveled by sun, wind, and age. Her expression tended to be severe, but behind her owlish glasses, her bright gray eyes missed little and often sparkled with humor. We made plans for Mary's visit. She'd go to Gibb's Farm after her "work" safari and relax with our mutual friends, Margaret Gibb and Per Kullander. David would collect her and bring her to Mangola.

The plan worked. David brought Mary with the usual strong wind and the last of the cool September weather. She climbed down from the Land Rover well dusted and tired. Sipping my welcoming mint and lemongrass tea, she told us a bit about the safari. Mary lamented about how few of the guests prepared adequately for the trip.

"I suffered the fools for days. None of them seemed sincerely interested. They knew little to nothing about rock art, let alone African rock art. By the time we got to Tarangire Lodge, in the end, I knew I couldn't spend another evening with them. I made the driver take me straight on to Karatu."

Mary paused, took another drink of tea, then chuckled. "When I got to Gibb's Farm, I was completely flattened. I could hardly talk. Margaret became my angel; she gave me a whiskey. Then another. I just sat there like a lump. On my third whiskey, Margaret suggested a bath. Ah, heaven! A bowl of soup and I immediately fell dead in bed." She snorted a sharp little laugh.

After lunch, both Mary and I took time out for a siesta. David, who seldom took naps, spent the time working under the Land Rover. We wanted the car to be in good shape for the short forays with Mary I'd devised. Ah, yes, I'd not yet worked up the courage to invite her on another day of bad roads and bright sun. But I only had this chance, so I determined I would somehow put on a bold face and propose local trips.

The next day Mary contented herself by reading, sitting happily in her director's chair under the tamarind tree at the guesthouse. David and I rushed around doing home chores, sneaking glimpses at Mary to make sure she looked content. At breakfast the next day, Mary said, "Would you believe it? I've already run out of good books." She pointed to a small stack of books on the table. "I've skimmed all these; they aren't worth reading."

"I figured you might run out, so here are some more to choose from," I said, setting a pile of books on the table. Mary leafed through the books while I stood watching, wondering how I could discreetly put forth the idea of a local safari. Finally, she looked up at me with a raised eyebrow. "Now, what are you scheming, Jeannette?" she asked.

I felt surprised she'd caught me out. "Well, umm, I wonder if you would like a little adventure in the afternoon."

"What adventure, pray tell?"

"Ah, a short trip over to Mumba Hole?" I suggested.

Mary at Mumba Hole

She smiled, already a conspirator, and said, "Mumba Hole, eh?"

Mumba Hole was a rock shelter site near the edge of Lake Eyasi, used by stone-age people for over 60,000 years. It had been excavated by a German, Ludwig Kohl-Larsen, and his Norwegian wife, Margit. They first described the site in the 1930s while searching for artifacts. Living in a tent they dug up the place. Margit oversaw the excavations. Our friend Per Kullander at Gibb's Farm was a distant relation of Margit. He showed me old maps tracing Margit's wanderings. She was a tough old gal, nearly 100 years old when she died. Lucky for us and others, she left behind accurate, precise records. Those lured other archaeologists to the site.

CHAPTER 9: TREASURE HUNTING

Mary and Louis were among the researchers interested in the Kohl-Larsens' reports. They visited Mumba Hole but didn't feel tempted to work there. An American named Mike Mehlman did. He came to reopen the site in the 1970s and 80s.

We met Mike in the 1980s. He'd discovered some interesting things. The results of a geochemical analysis on the sharp-edged obsidian tools were a big surprise. The obsidian came from around Lake Naivasha in Kenya, 230 kilometers to the north. He concluded that the stone age people had extensive trade networks in East Africa.

Another of Mike's exciting findings was the considerable number of snail shells found at living sites around Mumba Hole. Presumably, people living along the shores of Lake Eyasi thousands of years ago slurped up snails. They foraged for food, painted on rocks, and traded with neighbors up and down the Great Rift Valley.

Mary didn't comment on these facts I offered about the rock site. She listened quietly then cocked her head and said, "Yes, I'd like to go back to see what's happened to Mumba Hole. Indeed, I think I'll be ready for a little adventure this afternoon."

I was glad she wanted to go, despite the significant disadvantage to the destination—getting there. We'd have to cross through numerous small onion plots, over irrigation ditches and fields. David agreed to drive, and my aversions diminished—somewhat. Being a passenger was often worse than hanging on to the steering wheel. We took Pascal along because he could help push the car when we got stuck. Also, since our Hadza friend Gudo was around, we took him along, too. He'd always been keen to see old rock art sites and had worked with Mary before.

We followed the main track through the onion fields. I counted the crossings: 17 irrigation ditches, one river, and one swamp. The trail ended at a big irrigation ditch. We would have to cross through flooded fields to get to the painted rock shelters on the far side. We parked and walked, wading through the irrigation water and across the lumpy hand-plowed fields. During the trek, we accumulated the inevitable followers: children and Iraqw men in ragged clothes. Goopy clods of mud clung to our feet. I don't do elegant expeditions, I thought.

When we reached the Mumba rock shelter, David and Gudo went exploring. I sat on a heap of excavated soil with Mary, her pant legs rolled up, feet muddy, smoking her cigar. She frowned at the deep hole under the overhang.

"Those inept archaeologists ruined this site! Why couldn't they take more care when they dug here? Look at all this debris they left. Anything interesting or valuable has been stolen or destroyed. The place is useless now."

While listening to Mary, I looked less at the devastation than at the rock art. The paintings seemed better than I remembered. One, with some 20 people all in a row, like cutout dolls holding hands, was still visible. Maybe the deep, unfilled hole deterred people from getting close and defacing the designs.

Mary and I talked over the idea that the descendants of the rock painters could still be around. Both the click-speaking Hadza and Sandawe were distantly related genetic groups that had lived in the area for thousands of years. We tossed around the idea of taking some Hadza with us on a safari to the famous rock paintings of Kolo, in Sandawe land. Mary had written a book about them, as well as revisiting them recently with her safari group.

We liked the idea and started making more definite plans when David and Gudo rejoined. They brought us a hyrax skull, some antelope bones, and an entourage of boys. By now, a crowd of onlookers surrounded us; we decided to leave. When we reached the car, the heat had not wholly melted the oatmeal chocolate chip cookies I'd made. We headed west in search of somewhere shady and quiet to enjoy a break and the scenery.

Six more ditches tried to foil us then the track wholly disappeared. Finally, we reached a spot on a hillock overlooking the vast expanse of dry Lake Eyasi. The landscape stretched out like some fantastic tawny lion pelt anchored on the far side by the purple rift valley wall. We ate cookies and sipped tea, watching the sun go down behind the rift escarpment in a fuzzy golden haze.

Mary sat quietly on a grass clump then said, "This is very beautiful. It's still wild here. I love the expanse of sky. And ah, the highland mountains are magnificent. What a view!"

Her facing Oldeani mountain seemed an opportune moment to bring up my next quest. "Mary, how about another little expedition? Would you be interested in searching for the Olpiro ruins around Oldeani? The ones you explored years ago?"

Mary looked at me and nodded. Her expression told me she knew I'd set a trap in a beautiful spot just to catch her in a good mood. She smiled, "Well, yes, why not?"

The two of us set off the next morning with a picnic. Our goal: to find old irrigation ditches and platforms made by people hundreds of years ago. Neither Mary nor I thought we had much chance of finding the site because of the happy regrowth of brush in the area. Curiosity and our love of adventure urged us on during the long bumpy drive to the north. After some sidetracks, we drove to the village school to ask for directions. It was already midday. I was worried Mary was hungry and bored with the safari, though she hadn't complained once.

"Would you like to stop somewhere here in the shade and have lunch?" I asked. "It could be our last chance to have a quiet bite before we check at the school. We might even have to take a guide with us."

"No need," she replied firmly. "I'm fine, carry on."

I steered into the school drive. Mary waited in the car while I went to ask the new headteacher about the road to Olpiro and the ruins.

"Karibu sana," said the headteacher. *You're very welcome.* He led me into his office, went round to his desk, and sat down in a formal pose. He pushed his guest register towards me. "First, please sign our book."

I sighed. Such formalities were necessary here in the outback of Tanzania. The delays always reminded me to try harder to be patient and endure with good grace.

CHAPTER 9: TREASURE HUNTING

I signed, forced a smile, and explained in my simple Swahili where we wanted to go.

"There are some old rock ruins near Olpiro village. They're near a place where there is water. We think the ruins are by springs or streams flowing from the rift wall. A long time ago people made irrigation ditches there and lined them with rocks. Perhaps you've seen them?"

The teacher surprised me with his knowledge when he said, "Hmm, I've seen ruins like that in the Manyara Rift Valley at Engaruka. Are there similar ruins here in Mangola?"

"Well, yes, some foreigners have come here in the past to look at the ruins, but no one recently. Do you know anyone in the village who could guide us?"

"If you need someone right now, I will find you a guide who knows the area. However, this man might not know the place of old stones."

I sat in the stuffy little office and waited. Back he came with a young fellow. We introduced ourselves. I thanked the headteacher and went outside, expecting our new guide to follow me. Instead, he glided away into the crowd of children and came back with his younger brother. Yes, we had to take him along too. A guide couldn't go out alone. Maybe two guides could better protect themselves from two aged white women?

Amused, I ushered them both into the back seat of the car. The Brothers sat, looking expectant. We set out again along the track, eventually finding the semblance of a road that passed the ranger hut that marked the boundary of the Ngorongoro Conservation Area. I stopped, looked around but no one appeared. Feeling like trespassers, I drove on.

I confess I have a significant failing; I am one of those bears who inherited the urge to go over the mountain. I always needed to see what is on the other side. Not always a good idea. Soon the bushy growth of thorny young trees enveloped us, screeching and scratching the car. I picked up signs of an old, eroded track and glanced at Mary. She sat stoically scanning the thick brush. The Brothers sat quietly in the back, obviously well outside the range of their local experience.

I was glad to see three people striding towards us. One was a colorfully clad Datoga elder with a white cap and red ear tassels. Walking alongside him was a short, skinny, black-robed youth. Further back came a black-robed young woman. I slowed to let them pass so we wouldn't cover them in our dust. I leaned out to greet them, then asked in Swahili,

"Do you live around here?"

"No, Mama Simba, we live at Kisima, in the palm trees by the lake. It is not far from your place in Gorofani. We met you at the bung'eda there."

As always, I was surprised, embarrassed, and a little pleased to be recognized. Looking at the man more closely, I asked his name. He replied, "I am called Jerumani." What a curious name for a Datoga I thought.

"Jerumani," I said, "Do you know about any ruins, rock heaps, or ditches built long ago?"

"Yes, there is a place by the hill over there. See that big gulley coming down the rift?" He waved at the blue wall stretched in front of us and added, "I remember some *Wajapan* [Japanese] who came to look at that place. They came to dig up old things, treasure."

It puzzled me that the Japanese seemed a nameable tribe and not lumped in with Wazungu, the collective Swahili term meaning pale-skinned foreigners. The Japanese seemed to have investigated every nook and corner of Mangola.

"Would you be willing to show us the place?" I asked.

"*Ndiyo, Mama*," *Yes*. He added that it might be too far to get there in what remained of the day. I should have taken his implied warning. Jerumani got in the car; his companions wanted to keep walking. We pushed on, crossing deeper and deeper gullies, virtual canyons. The road became a cattle path; the patches of thorn bush got thicker.

Mary sat quietly, holding tight to the dashboard or door handle as we bumped through the bush. I pulled the car this way and that to avoid the worst of the thorny trees, a stump, or a thicket. The midday heat stewed us inside our solar cooker of a car. I projected onto her my internal state—frustration, thirst, hunger, and an overwhelming wish to stop jouncing about in the car. Even inquisitive bears need to eat and drink. I gave up when the brush grew so thick that we would need to hack it away to pass. Not prepared to bushwhack, I admitted defeat and turned back.

Thorny branches poked like ghoul's fingers through the open windows. Tsetse flies came, too; they were at their worst, overactive in the heat, stabbing ankles and necks, drawing blood and cries of Yooch! We had to close the windows. Right away, the waves of powerful body odor washed over us. I could almost feel Mary longing for a cigar to fight back the smells and deter the insidious flies. She resisted, she later told me with her twinkle, because she thought our guides would ask for a cigar too.

I handed back water bottles to the fellows in the back seat. They'd have to suffer and smother a little longer inside our oven. Not a word was spoken as we headed back towards the village. To have our safari end in a let-down discouraged me. I mumbled, "What a goych," under my breath. Mary asked, "OK, Jeannette, tell me again what 'goych' means."

"Goych means unfinished, frustrated, left hanging. It's a useful word for a time like now. We had expectations, hoping to find something. It just didn't happen. We'd set out to find the ruins, made a real effort, but we're goyched by ignorance and terrain."

Mary nodded and chuckled, "Well, that's a useful word, and though we've been goyched, it's even worse." She paused, turned to stare at me, and said in a low voice, obviously trying to keep a straight face, "You've forever ruined your reputation as a daring bush trekker by turning back."

I laughed, she snorted. She added, "Actually, I'm delighted we're heading home. I won't tell anybody we never made it to the ruins. Louis and I saw them many years ago. That's quite good enough for me. I'm rather glad they are buried in the bush now; it might protect them a little longer."

Slowly we backtracked towards the Conservation border and ranger post. I began to worry about how we could get back into the village without seeing the rangers. We'd been roaming unchecked for the past several hours. And of course, there they stood in the middle of the road, two uniformed, intimidating Ngorongoro guards, their faces set in official scowls.

One held up his hand and stomped up to my window. Thrusting his face close as I tried to roll down the window, he declared, "You must come to our office now. You are in the area illegally; you must register at once."

I started to explain who we were. With head high, the ranger turned away and marched to the side of the road and into the office, demanding, "Come to the office."

The other ranger seemed embarrassed by his brusque partner. He tried to appease us with reassurances that they only wanted us to sign in. Once in their compound, the snarly ranger demanded we reveal ourselves. I meekly said, "I am Mama Simba from Gorofani village, and this is my guest, Dr. Mary Leakey. She works at Olduvai Gorge and Laetoli; she is an archaeologist for the Conservation Area."

Their abrupt change in demeanor surprised me. Suddenly friendly smiles appeared. "Oh, yes, we know who you are. Welcome. Please sign the guest book; we are glad to see you."

With great relief, we shook hands all around. Mary and I signed the book, thanked the rangers for their welcome, and went back to the car. Jerumani had walked on, but the Brothers sat stiffly, waiting. At the school, the Brothers left us, shaking our hands, accepting my oranges and cookies. I took the time to thank the headteacher properly, attempting to show the polite manners expected by Tanzanians.

On leaving the office, I found Mary surrounded by Hadza children who turned to me and called out greetings: "Mama Simba, Mama Simba, Shayamo." I greeted them in return and bowed. They all giggled. I stood there, bemused by the scene. Singing came from a classroom where a young fellow sat on an open windowsill that had never seen shutters or glass. He was playing guitar and the Hadza children danced around, singing. I wanted to delay our leaving and enjoy the moment, so I went to get cups of water. Mary sat on a log and listened to the songs as though the intended audience of an impromptu performance.

We set off, but I had one last plan. I pulled off the track and parked under a large thorn tree. From the back of the car, I pulled out two folding chairs and a little table. I made an artful flourish with the kanga cloth as I covered it then started laying out our long-delayed picnic.

"Jeannette! You're a magician!" Mary exclaimed and sat down gratefully. The water in our flasks was still hot enough to make tea, and we ate with gusto. Mary lit her longed-for cigar and smoked while we lounged in our chairs. Farmers and herders greeted us as they passed by with donkeys, sacks, and tools. No one stopped to stare or ask for anything.

We sat in the gentle gauzy light. Oldeani Mountain loomed to the north behind us, cloaked in a thick rust-colored haze. A similar pall of dust almost obscured the rift wall to the west. Mangola was usually a dust-covered place but now I wondered if some of the haze came Mount Pinatubo in the Philippines. Its eruption in June of the previous year had been the second-largest of the century. Over the months, Pinatubo's outfall circled the globe. I was reminded that even though we lived in a remote corner of the world, we could and would be affected by what happened elsewhere.

Mary sighed and said, "This feels like the old days when we went exploring in wild places. It feels good to be far away from the busy cities and demands. I'll be having to cope with them soon enough. For now, rustling palms, fuzzy dusty light, and another night or two in the wilds of Mangola feels fine. Ruins or no, it's been a most enjoyable safari. Also, we can look forward to planning our Kolo Rock Art safari. That is guaranteed not to be—as you call it—a goych."

From our Mikwajuni guestbook

CHAPTER 10
SINGING STONES
A DISCOVERY SAFARI INTO YAEDA VALLEY AND HILLS: 1992

Kampala

"Gudo," I said in Swahili, "will you go with me to look for Kampala?" We sat together in the kitchen bottling jars of honey. Most of the morning we had been straining raw honey from the hives we'd strung on our sturdiest tamarind trees. It was a sticky business.

Gudo looked up from this task, puzzled. I realized that Kampala could mean not only the person but also a big city in Uganda.

I smiled and nodded. "Hadza Kampala," I said. "I think he'd be a good person to come with us on our safari to the Kolo rock paintings. He always has lots of ideas. He's still living in the Yaeda Valley, or so I heard. Do you think he'd be an OK member of the team?"

Gudo poured honey through a filter into a jug, thinking. I wondered if he worried about Kampala as a team member. Or, more likely, worried about having to search for Kampala in the Yaeda Valley, a far away and troubled place. A group of Hadza lived there in a cluster of huts that had been registered as a village. The village was called Mongo wa Mono. Government bureaucrats and concerned outsiders had enticed or forced the Hadza there, depending on your point of view. In either case, the intent was to make the nomadic Hadza settle down. According to the propaganda, they would be better off, with access to water, schools, clinics, and employment.

But most Hadza didn't want to settle down or live in a village. They were bush people. They lived in small camps. When living close together in a village-like situation they started to wilt and die. They came down with illnesses and succumbed to manipulations by politicians, do-gooders, missionaries, and tourists.

If Kampala lived there, we'd have to go to Mongo wa Mono. There was no other easy way to reach him and propose our rock art safari. Kampala had lived near us when David and I first came to Mangola. We knew him as a lively, loquacious, and imaginative man whom we liked very much. He was one of the three men I wanted to join the safari Mary Leakey and I were planning.

As well as finding Kampala, I also wanted to check out the conditions in the Yaeda Valley that the Hadza complained about. But I was hesitant because Gudo and I most definitely would be crossing paths with touchy, self-important politicians, or xenophobic local administrators. Even so, we had a purpose. I sighed and asked again, "Will you go with me to Yaeda Chini?"

Gudo tilted his head, looking at me with hidden mirth. He told me slowly and carefully as though teaching a child, "Mama Simba, if we go to Mongo wa Mono village, maybe we will find Kampala. Good. But maybe we see Naftal and political people who cause us problems. But also, we can escape into the hills." Gudo broke into a grin adding, "We can go to the gong rocks!"

Wow! was my first response. Yes! was my second. I'd been wanting to search for the gong rocks for a long time. Archeologists and the Hadza had told many tales of painted rocks and singing stones in the area. I especially wanted to see those gongs. Gudo nodded, pleased that he'd piqued my interest.

I poured clean honey into a jar and put on a label mindlessly, already imagining the adventure into the wild bushland of the Yaeda Hills. The mysterious singing stones were hidden somewhere well off the beaten track, no roads or trails to follow. There would be no helpers, no way to communicate with the outside world. The thought of these marvelous challenges started my mental exploration motor. My enthusiasm for the trip roared into life. I said, "Yes, Gudo. Good idea, we'll go on a double quest, Kampala first, then rock gongs."

East Africa, especially the Lake Victoria region, has many granitic rock outcrops. Humans pounding on specific boulders for centuries created cuplike indentations. These are the gong rocks. They ring out clear notes when struck with a stone. No one seemed to know much more about them. Now I had a chance to see one in Hadzaland!

David and I first discovered gong rocks among the outcrops of the Serengeti

CHAPTER 10: SINGING STONES

plains, then at places around Lake Victoria. Modern people thought they'd been used over the ages to communicate. We disagreed. David and I tried walking various distances from friends bonging on the rocks. We knew that the sound didn't carry far. We figured that ancient people used to gongs to make music together. We imagined them in the moonlight, singing and dancing to the rhythms.

On the morning of our Gong Rock Safari, four of us set out in good cheer—Mama Simba, Gudo, and Sagiro—the other member of the proposed Hadza team. The fourth was Pascal, our Man Friday, who could be a help setting up camp, driving, and fixing the inevitable puncture.

All of us loved an adventure except Pascal. He'd agreed to go because I needed him and he got to drive the Land Rover. Once in the driver's seat, Pascal put the car into gear and his face into his worry frown mode. Finally, he began to relax when we crossed the bridge on the far side of our village. Up and up we went, into the rough hills and across braided dry streambeds. Finally, Pascal turned off the rutted track into the canyons that led to the eastern-most end of the Yaeda Valley. We followed a dry river course until we came to a cluster of baobab trees. The lovely wild spot claimed itself as our base for the night.

Gudo and Sagiro vanished into the bush to look for their fellow Hadza hunters. They soon returned. "There is a camp here, not too far," Gudo told me. He motioned for me to follow. To me, it was magic how the Hadza found each other in the bush. Familiarity with old camps was part of it, but even when turned around and coming from another side, in another season, in another year, they could find each other. This time we walked into a camp of about five Hadza families hidden in the bushes. Dusk had already settled on the landscape, so I peered around and greeted the silhouettes sitting here and there.

Men, women, and children sat on the ground. Even in the dim light, I could see the remains of their supper—hard, sweet *grewia* seeds and sticky *kongolobi* fruits mounded in bowls made from the shells of baobab fruits. Everyone had already stuffed themselves, so sat around chatting.

We agreed we'd return in the morning for more news and talk.

After supper back in our camp, Gudo and Sagiro sat smoking their pipes by the fire. Pascal busied himself with clearing things out of the way of roaming hyenas and jackals. I gratefully escaped to my usual preferred sleeping quarters, the mattress on the roof rack of the Land Rover. A thick band of stars blanketed me as I fell asleep listening to the distant coughing purr of a leopard somewhere.

Waking at dawn, I rolled up my bedding, climbed down, and started the fire. I filled Captain Kettle with water and put him on the flames to boil. The men emerged when I'd made enough racket to rouse them. We ate some peanut butter and honey sandwiches, the men washing it down with great cups of sugared tea. While Pascal started breaking camp, Gudo, Sagiro, and I went back to visit the local Hadza enclave. I carried my usual visiting kit—gallon jugs of clean water, supplies of matches, bandages, eye medicine, and rehydration salts.

Gudo negotiated for arrow poison while I did some cleaning of wounds and infected eyes. Except for some burns and sores, the group glowed with health. They had fruits and meat in abundance, as well as freedom. Their only complaints were worries that officials might round them up and force them to go live in the dreaded Mongo wa Mono.

Back at our baobab camp, Pascal loaded the rest of our gear. He looked more irritated than usual, so I gave him the job of co-pilot instead of driver. If he drove, he would just get more frustrated. We set out cross-country around the end of the dried-out swamp. I forced the Land Rover through the thorny bush, trying to follow cattle and people tracks. At last, we came to a big sandy river. The banks of the dry wash had a border of bushes laden with ripe *madabi* fruits. Irresistible! I hopped out and climbed on top of the Land Rover to gather in the sticky persimmon-like berries from the high branches. Gudo and Sagiro waded into the bushes to strip off the lower limbs, putting some in the car. I wanted to take the berry branches as a gift to the Hadza trapped in the village—the berries were Hadza comfort food. We gorged ourselves as we drove onwards, my steering wheel getting stickier and sticker from the gooey fruit.

Cutting through thick scrub, we encountered another surprise group of Hadza. Three hunters popped out of the bush, waving their bows and arrows at us. We stopped to greet and discuss the weather, Hadza style. Did you find the best berry patches? Any animals to hunt? Seen any beehives lately? They gave us a bag of baobab seeds for people in the village of Mongo wa Mono, as though sending food to prisoners.

We left the hunters behind and kept grinding along until we came upon a sort of tunnel through the thick thorny bushes. Without warning, two donkeys sauntered into the track ahead. They simply would not get out of the way despite my honking and coming close to their rear ends. Instead, they trotted ahead of our car, and I couldn't get by.

I slowed in hopes the donkeys would move aside. Alas, they also slowed, then stopped to nibble at the bushes along the track. I revved the engine, but the donkeys only ambled away, still blocking us. We all laughed, but I got annoyed. To pass them, I bounced through thorns and rejoined the road ahead of them. We all cheered as we roared away, leaving the donkeys far behind.

Mongo wa Mono came into view, a depressing sight. Some huts stood here and there, spread out around a bluff of rocks like pimples around a big nose. The barren ground glared at us, reflecting the hot afternoon sunshine. A large pile of stones, gravel, and mud bricks sat near the road, clearly building materials.

I headed for a spindly tree by the bluff. We stopped in the scanty shade and scanned around. At first, we couldn't see anyone in the heat shimmer, and our eyes stung from the glare and dust swirls. The most arresting thing in the dismal landscape was a massive flesh-colored baobab tree. It stood like Gulliver, a giant in the bare expanse, with log beehives hanging from its stout branches.

CHAPTER 10: SINGING STONES

We got out of the car, staring around the place. I took a picture of the baobab with hives. Just then, the donkeys arrived, trotting right up to the vehicle. I laughed and turned to Gudo to point out that our four-footed friends had returned. Gudo was looking the other way. He did not smile.

"We should leave," he said in a low voice. "Right now." He gestured with his head to a delegation approaching from the cluster of huts. I knew we were already trapped. It would be rude to flee from a greeting commission. Also, we hadn't yet found Kampala. I tried my best to look composed even though I was sweating in the fierce sun, tense and tired.

I started a distraction ploy, getting the branches of madabi berries from the car, as the reception committee assembled. Squaring my shoulders and summoning sociality, I greeted the leader. "Hujambo," I said. "Sijambo," he returned. We looked at each other. I recognized him. He was Naftal, a tall man for a Hadza, his stance that of someone who considered himself in control.

I'd met him only briefly before, but we knew each other by reputation. Naftal had appointed himself chairman, if not dictator, of the Hadza tribe. He definitely was in charge of those who lived in the new village of Mongo wa Mono. I was but a lowly Hadza Friend who lived in the infamous Mangola area where the decadent Hadza lived.

While we assessed one another, we conversed casually in Swahili about roads and rain and other innocuous subjects. Then Naftal started to ask me questions about researchers and politics, and why I'd come to "his" village. I told Naftal that I'd come to see Kampala and asked where he might be. Naftal claimed not to know, but I sensed he wanted to know why I was looking for him.

I tried to change the conversation by giving the Hadza the offerings of the berry branches and baobab seeds. Gudo and had cleverly gone to the other side of the car to dole out the madabi berries, thereby avoiding a confrontation.

Naftal continued asking me questions about Mangola while the afternoon sun dazzled me. My mouth and throat were dry. There was nowhere to sit, no shade; I was literally being grilled. Two missionary-type Wazungu appeared and made me even more uncomfortable. We chatted with much posturing and smiling. Finally, I felt I could bear no more of what Gudo called "blah-blah." I had to leave.

Appropriately, one of the donkeys pooped right behind Naftali and distracted us all. The children picked up rocks to throw at the beasts. I grabbed the chance to tell the congregation how pleased we were to see the famous village of Mongo wa Mono. Thank you, but now we had to look for Kampala before dark.

We drove through the scattered huts towards a bunch of Hadza standing at a hand mill, grinding maize. They broke off to greet us. When we asked where Kampala might be, they pointed in a vague direction. Gudo and Pascal climbed out of the hot car and went to look for him.

I parked under the skeleton of a tree and drank some water. Sagiro looked at me, then the surroundings. He said in a slow, sad voice, "This is not a good place for people." I agreed, holding my head, trying to keep my heat headache from blowing my brains asunder.

As I was about to abandon hope, Gudo and Pascal arrived with Kampala. We smiled hellos at one another then negotiated the details of our intended safari. Kampala agreed to come along on the trip. With a big smile, he said with enthusiasm, "You don't need to come here to get me. I will come to Mangola!"

Volunteering for some days of walking meant he really did want to come along. Also, of course, he would get to mingle with other Hadza all along the way. We waved goodbye with great relief and departed. I drove around the back of the bluff to avoid meeting Naftal again. Welcome stretches of bushland decked with baobabs and acacias swallowed us up. At last, I felt safe enough to pull off the road. As we got out of the car, stretched, and started to set up camp, Gudo gave a hoot of laughter. He pointed up into a tree where an owl perched, staring at us.

"Mama Simba, look! Its face looks like Naftali when he saw the donkey poop!"

Spotted Eagle Owl

CHAPTER 10: SINGING STONES

We all laughed and laughed, so glad to get away from that horrible village. Gudo and Sagiro went to bed right after we ate, but Pascal felt tense. He wanted a cup of coffee. I tried my best to dissuade him, knowing that he was high-strung anyway. "Pascal that's not a good idea. You know that Tanzanian instant coffee is tasty, but it's very strong, it will keep you awake if you drink it now." His response was, "I'll put lots of sugar in it," as though sweetener would somehow mitigate the effect of the caffeine.

He insisted, and, with misgivings, I fetched it for him. I, too, was feeling wound up, so crept off to a platform made by the twisted roots of a baobab. I played my *filimbi*—my little flute—and looked up at the gorgeous spray of stars. Finally, I felt soothed enough to go to bed. From my car-top mattress, I saw Pascal still sitting by the fire. He looked altogether too wide-awake; he probably wouldn't sleep a wink all night.

In the morning, I watched as Pascal pulled himself out of his tent, looking worn out and agitated. His gloomy mood didn't deter the rest of us from enjoying the journey away from Mongo wa Mono into the lush Yaeda Chini Valley. In the past, the area was the preserve of Hadza hunters, home to large herds of zebras and gazelles. Now those animals had gone, replaced by Datoga settlements, cattle, and goats.

Tsetse fly

We drove around the far side of the dried-out swamps and followed a little track into the Yaeda Hills. Somewhere ahead, we hoped to find the gong rocks. The arch defenders of wilderness—tsetse flies—soon welcomed us. Their painful stabs were the price we paid for intact wild spaces. Tsetse flies sometimes carried sleeping sickness, and, in infested areas, cattle and people could die from the disease. Thus, pastoralists avoided tsetse areas, and wildlife could remain. Among the more significant members of the bushland greeting committee were impalas, giraffes, and kudus, all naturally immune to tsetse-borne diseases.

I was humming an I-am-joyous-while-roaming tune as we entered the fly zone. The Land Rover joined in with a low mutter as we ground along a well-used track winding upwards through bush and baobabs. We followed the track to the crest of the hills that divided Yaeda from Lake Eyasi. Just before the top, I noticed a side road. I stopped the car and asked, "Which way?"

"Not that way," Gudo said and pointed to another track going over the ridge. I looked at the vague road that dipped into a rocky valley, but Sagiro said, "Not that way."

"Well, my friends, tell me which way," I said, amused.

"Let's go back. I think the track is further back." Gudo said.

"No, it's that way, over there, ahead." Sagiro gestured to a vague unused track probably made by elephants.

"I wouldn't trust either direction," grumbled Pascal.

They all started arguing.

"OK guys," I said, "Which way? You decide, then we go there." Gudo and Sagiro gestured this way and that. Pascal rolled his eyes.

I finally said, "Gudo, can we drive nearer the rocks, or should we park and walk from here?"

"No, Mama," he replied. "We are still far. We must get closer. But from here, I do not see the right road."

I looked at Gudo fondly. I trusted him. Somehow, he'd get us to the gong rocks. Suddenly I was awash with that light-hearted emotion that I hope you've shared, the feeling that you don't care what happens as long as you are in a beautiful place, have good companions, and enough resources with you. I felt thrilled to be alive, my inner self in neutral, happily awaiting whatever Gudo and Sagiro decided.

I pulled the Land Rover off the track into a bit of shade. Pascal and I sat under the scrawny acacia tree, eating bananas while Gudo and Sagiro wandered off into the grey-green bushes. They came back with two Hadza boys. Gudo's skill at finding Hadza in the bush amazed me once again. The boys offered me some delicious small-bee honey wrapped in leaves. We hit it off immediately. My delight increased when it became clear that they knew where we wanted to go. They pointed down the trace of a track. I drove carefully with much bumping and crashing, hoping David wouldn't notice all the new scratches on the car.

Our new guides pointed us back down the slope of the hill and onto another track. A rocky hill marked the end of any possible further crashing and bashing. We piled out of the car. I shrugged into my commando role.

I gave Hadza Boy One the job of perching on the roof rack. He would look after the car and call to us if we didn't return before dark. Boy One seemed pleased with this task, especially since I gave him a bag of peanut butter and honey sandwiches.

The rest of us followed Hadza Boy Two, Sagiro striding confidently out in the lead. Gudo shouldered a pack with water and fruit, and I toted my bag with cookies, a camera, and binoculars. We set off across hill and dale and swamp, crossing a muddy area guarded by baobabs. Animal tracks in the soft earth surrounded a small pool of water.

We walked around and through thorn thickets. When we could, we bounded over rocks set between thickets and grass. The burnt grass

Cycad plant

Gudo and boy at Kidero gong rock

was especially prickly, the stubble poking into our feet.

We paused to look around.

Sagiro admitted, "The gong rock, hmm, I can't remember exactly. I know it's somewhere around here." We carried on to a high spot nearby. Gong rocks, I reasoned, often sit on top of the biggest rock outcrops. Sagiro headed off into a nearby gorge. I dumped my pack next to Hadza Boy Two and headed to the top of the rock outcrop with my binoculars.

I scanned around and got a surprise—cycads! The stunted palm-like plants grew in clumps, crammed into crevices in the rocks. I wanted to go back for my camera to photograph the rare plants, but heard a shout echoing out of the gorge ahead.

I slid down the rocks to Boy Two. Sagiro and Gudo stood high on a granite slope, waving us over to their side of the gorge. Boy Two and I scrambled through the rocky cleft to the other outcrop. Reaching the top, I gazed at the stark and awe-inspiring setting.

To the north, the Ngorongoro highland mountains etched a broad purple massif against the dusty blue sky. To the west, the fuzzy blue Eyasi rift wall bordered the shimmering dry white lakebed of Eyasi. We stood among the ragged Yaeda Hills, broken and wild. And right in front of me were not one but three singing stones.

The three rock slices, like wedges of fossilized cheese, faced eastwards onto a valley with more rocky hills in the background. Each rock slice had the cup-shaped holes made by humans hitting at them but showed no recent signs of use. I picked up a stone and banged at the hollows. To hear the varied tones from the different dents made me laugh.

We played around with the rock gongs until the music in us dried up in the hot sun. Retreating from the bright bare rock to a patch of shade, we sat and ate the tangerines and cookies. We lay down like lizards on the smooth granite and enjoyed the vast peace that pervaded the place. Ah, the quiet, with just the gentle breeze brushing sweat bees from our faces.

My imagination filled with images of ancient peoples using this special place. Here they could sing and dance, celebrate, have fun, rest, and enjoy life. The aura of the place lifted my spirits, soothing my soul. I fell into a dozy trance. Gudo gently roused me, and I sat up alert, realizing we needed to get going if we wanted to get back to the car before dark.

We crossed the open granite faces, keeping on top as long as we could. When we ran out of high spots, we bounced down the rocks to cross yet another gorge. I felt like a transient molecule in the landscape, dancing through time, living a short precious lifetime. Here I was, my genes shared with all living things—the cycads, the sweat bees, and my human companions. I felt connected; I felt good. Just writing this brings back those feelings of freedom and joy.

Alas, my effervescent feelings evaporated as the hot afternoon wore on. Water and food awaited us at the car, but it seemed a long way off. Walking became more and more difficult because my rubber tire shoes cut into my little toes, and the gritty soil had rubbed blisters on the soles of my sweaty feet. We reached a wide valley where smoldering or burnt grass stretched like a begrimed carpet. I plodded along, trying to step between patches of stubble.

But I couldn't miss them all. The short stems of the burnt grass were hard and pointed. One jammed into my right foot and sliced it open. I slowed down, stumbling, dropping behind the rest of the gang. I lost them as they forged ahead. I wasn't worried about getting lost in this wilderness—clear landmarks abounded—but losing my buddies perturbed me. I shouted out and at last caught up to Boy Two and Sagiro, who had waited for me. I smiled my thanks and asked where Gudo and Pascal had gone. They waved in the direction of some hills.

We three continued, going through still-burning grass clumps. I wondered who'd started the fire and why. My best guess was livestock herders. Eventually, we came back to the baobabs in the swampy place. I stopped to squidge my feet in the cool mud that both soothed and irritated my sores.

I looked down and found a welcome surprise—relatively fresh elephant footprints and dung. Those great beasts still lived here. Human prints were visible too, not barefoot either, recognizable flip-flops. I knew they belonged to Gudo. How he maneuvered through that landscape in them was as astounding as finding elephant tracks and cycad clumps.

Boy Two knew a shortcut back to the car. We arrived in the gloaming, the last glimmers of sunset painting the sky pastel shades of gold and pink. I washed my dirty feet, hoping Pascal and Gudo would arrive soon. Sagiro, the boys, and I hooted like baboons and chimps to help them locate us. Finally, over the hill, they came. We all climbed into the car, and I retraced our route in the dusk, finding a perfect campsite right before the main track. The Hadza boys took Gudo and Sagiro to their camp while Pascal and I started a fire.

Pascal looked tired and grumpy, but Gudo and Sagiro returned, jaunty, somehow rejuvenated by the long hot day. Together, we prepared a pot of rice and beans adding onions and all the rest of our tomatoes and sweet peppers. I sang and played my little flute to entertain them while the food cooked. After we ate, Pascal said he was exhausted, something he had never admitted before. He looked at me with a funny frown.

"I think I'll have a cup of coffee," he said, taking the Africafe out of the box. I looked at him wide-eyed. Was taciturn Pascal making a joke, or had he already forgotten the effect coffee had on him? I asked, "Do you really want coffee now?" He gave me a sideways look with what I thought might be a trace of a smile. He put the tin of coffee back.

We all slept well, lulled by the sweet singing of the Hadza in their camp nearby. I woke once or twice to appreciate the serenade of distant lion roars and hyena whoops. The sky twinkled and sparkled above me; what a splendid day and night. I tucked my good feelings inside me, thinking that at the moment, I could die completely happy.

We woke early with a crescent moon hanging in the baobab's limbs like a languid lover. We ate leftovers for breakfast. Three Hadza men and the two boys materialized from the brush as we finished eating. Luckily some food remained in the pans. They sat on the ground and eagerly ate the leavings. I doled out the remaining supplies of sugar, fat, beans, rice, maize flour, and onions. We kept our honey sandwiches, a bit of fruit, and our remaining water for the journey home.

We returned to the rocky track soon after the sun blasted over the horizon. Impala and a glimpse of a kudu were our early morning reward, but soon we left the land of tsetse flies and wildlife. We rejoined the modern world of Lake Eyasi, following cattle tracks along the lakeshore. We passed men with spears on their shoulders, herds of cattle, herd boys, goats, donkeys, people carrying wood and poles.

We reached home in the late afternoon. Leftover beans and rice made an acceptable post-safari meal. After putting his plate in the sink, Pascal pointedly made me a cup of coffee, an unusual gesture. I reckoned it was his way of making sure I got his jest of the night before. "None for you, Pascal?" I asked chuckling. "No," he said with a sly smile.

Gudo and Sagiro wrapped up their rewards of flashlights, bags of rice, and jars of honey in their new blankets and jovially strode off, Gudo still wearing his plastic sandals.

As I bathed my ruined feet, I marveled at how Gudo had walked—and gracefully—over that rough terrain in flimsy flip-flops. I imagined our human ancestors walking in bare feet for millennia. Their feet had stood on those granite rocks and pounded out rhythms. Were they lucky to live without houses, cars, cows, coffee, and flip-flops? My barefoot mind let the thought drift away and shuffled into sleep.

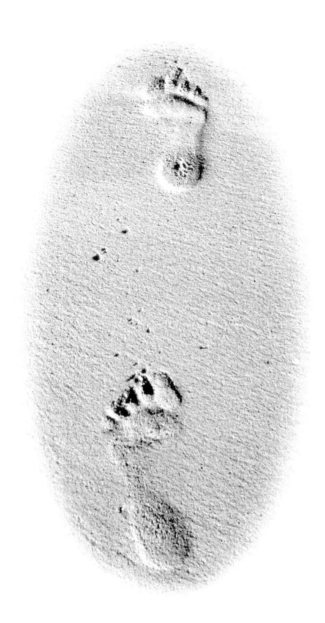

CHAPTER 11
KOLO ROCK ART SAFARI
ADVENTURING AND ROCK PAINTING MYSTERIES: 1993

Prehistoric rock art, Kolo 1

David and I were ready but where were the three Hadza men? We were still in Mangola while the other members of our safari awaited us at Gibbs Farm Lodge. Yes! our expedition to explore the Kolo rock paintings in central Tanzania was starting.

Mary Leakey and I had been planning this safari for months. Mary, our friend from our Serengeti days in the 70s, had remained an adventurous companion. Her profession as an archaeologist and her love of rock art made her The Expert in our little group.

The rest of us had roles as support cast and curious participants. Our expedition, besides pure pleasure, aimed to expose us and three present-day Hadza hunters to ancient paintings. I'd conscripted Gudo, Kampala, and Sagiro to join us. We all hoped they would have insights into the meaning of the designs. We'd share our interpretations, maybe even come to some conclusions.

Our friends Margaret and Per at Gibb's Farm had helped enormously. In addition to their lodge, they ran a safari business. They'd sent the head of Gibb's Farm

Tours, Nigel Perks, his assistant, Andrew, and staff to Kolo set up the campsite. Now, I felt pressured. Where were the three Hadza men?

My spirits rose as Gudo, Kampala, and Sagiro entered our compound. We stuffed them and ourselves into a car already stuffed with equipment, tools, and supplies. The heavily loaded Land Rover growled its way up the Horrid Road to Karatu.

At Gibb's Farm, we joined Margaret and Per, Mary, and Mary's granddaughter, Julia. From there we set out in two cars, a full day's drive ahead of us. From the highlands, we descended towards the Great Rift Valley floor. On the edge of the Manyara escarpment we stopped at the viewpoint. Everyone fled the hot vehicles to look out over the lake and woodlands of Lake Manyara National Park.

Gudo took my binoculars and the three Hadza passed them around, excitedly pointing out the buffaloes, elephants, giraffes, and many other animals they could see. Lucky for the animals, they lived in a protected area, escaping the Hadza. On this safari, the bows and arrows had been left behind by request.

The cars in low gear, we thundered down the rift wall. The Gibb's car went on ahead while we stopped at Mto wa Mbu, the village named after its nearby "river of mosquitoes." Kampala, Gudo, and I went to search out stems of eating and cooking bananas while David filled the car with diesel. Kampala immediately started to bargain with and tease the stall owners, acting like a buffoon who'd never seen a banana. Nearby shoppers found him hilarious, and we all laughed until our sides hurt. Finally, we bought three stems and fought our way back through the t-shirt and bead necklace vendors to the Land Rover.

Miles of dust and corrugations later, we reached the junction with the Great North Road. This route allegedly goes from Cape to Cairo, about 6,500 miles, the full length of Africa. Our portion of the drive would be but a hundred miles or so. We turned south and for a short while purred along, enjoying the smooth tarmac. Then the road deteriorated into a nightmare of potholes and broken asphalt.

After a couple of hours, we needed a break from the swerves and jouncing. Coming up a winding road onto a high ridge, we stopped for our picnic lunch. A colorful miombo woodland surrounded us. The miombo trees stood slender and delicate above open ground. They wore deciduous leaves of gold, ochre, amber, crimson, chartreuse, and lime. The palette was very different from our typical acacia woodland, dressed in subtle shades of green on green.

Settling ourselves on mats, we enjoyed the gourmet food Margaret had ordered from her team at Gibb's Farm. A bottle of crisp white wine made the picnic extra delicious. The Hadza men picked at the odd delights then wandered about crunching cookies, investigating the trees, and collecting bark for medicine. The rest of us talked about the history of the area.

The forested plateau where we sat, the rift wall, the waters, and the valley below had attracted ancient peoples who hunted and gathered there for thousands of years. We reckoned that the ancestors of the Sandawe or the Hadza had surely

CHAPTER 11: KOLO ROCK ART SAFARI

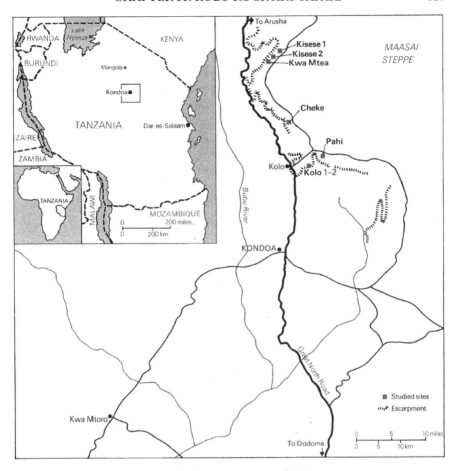

Map of rock-painting sites around Kolo

lived in the area. Those ancient peoples were the most likely to have made the rock paintings we planned to visit.

None of us knew much about the click-speaking Sandawe who'd inhabited the area. At present, they'd mostly given up their foraging lifestyle and settled down to farm alongside the Swahili. The Bantu people had drifted inland along with some Arabs from the coast during the time of the slave trade. We could still see evidence of that slave route in the ancient mango trees planted around watering places in the valley and at the base of the rift wall.

From our picnic site onwards, the road improved. We sped along, arriving suddenly at Kolo, a tiny village crouched alongside the way as though waiting for a ride. We pulled off in front of a peculiar A-framed structure made of stone. An old, badly cracked, Swahili-style carved door surprised us. Languages and new religions had followed the slave routes. The immigrants left traces that overlaid the original inhabitants, those who presumably made the rock art. The Zanzibar door was a clear reminder of the coastal influence.

Mary, Gudo, Sagiro, David and Kampala at Antiquities office, Kolo

The carved door opened to the Department of Antiquities' guard-post and office. A gangly man with a Muslim cap greeted us. "Welcome. We've been expecting you. Your camp crew alerted us of your visit. My name is Juma, chief guide. I'll be with you on your tour of the Kolo rock paintings."

We signed in, took photos of the odd building then left, eager to get to camp. Continuing eastwards on the side road, we saw tents set out in a sparsely grassed clearing. Glad to leave the cars, we let the busy camp crew help unload and stow gear in our tents. I noted that no shade trees grew near any tent. Midday would be hot.

Luckily, evening enveloped us quickly, bringing a welcome coolness. We assembled by the crackling campfire. With drinks in our hands and the delicious smells of dinner and woodsmoke swirling around us, we beamed at one another—a gaggle of adventurous friends in the promising light of a filling moon. I started humming to myself, a sure sign I was happy.

On our first day to explore the Kolo paintings, we woke to tents flapping in the cool dawn wind. One by one, we came to get coffee or tea. Mary, wrapped in a blanket, remarked, "Usually, I like the cool season, but here it's downright cold!" We huddled around a little campfire eating toast with homemade jams and lemon curd. Juma arrived with another guide named Pius.

I hung around a while but decided I didn't want to wait for the "primate pod" to assemble. David and I use that phrase to describe the amoeba-like way groups of primates get going, with many false starts like pseudopodia reaching out. Eventually, the whole organism moves off in a specific direction. Impatient to get off, I told David, "This primate pseudopod is now heading for the rift road. Please pick me up when the rest of the group finally gets going."

CHAPTER 11: KOLO ROCK ART SAFARI

Swinging along the gravel road by myself, I felt a special joy at the start of a new adventure. I'd gone about a mile when the first car caught me up. I jumped in, happy to bump along the washed-out track. We descended to a former campsite where Mary and Louis Leakey once stayed. Those two intrepid archeologists, with their assistant, Giuseppe della Giustina, had spent three months in 1951 tracing the major Kolo paintings.

Decades later, Mary's daughter-in-law, Meave, saw the tracings of the paintings at the Museum in Nairobi. She helped Mary put them together in a book called *Africa's Vanishing Art*. Once that book came out, more visitors came to see the Kolo sites.

Alas, visitors of all kinds brought problems to the rock art. Tourists threw water or even kerosene at the paintings to make them temporarily more vivid; bored goat-boys threw stones at the ancient art, and graffiti was common. Every day, the weather significantly affected the designs. The Department of Antiquities built cages of wood and wire to protect some of the sites, but the paintings steadily deteriorated with time and abuse. We felt glad we were still able to see some of them.

We left the cars with Pius and stood around admiring the granite rocks sparkling with mica flecks in the morning light. The nearest site, Kolo 2, clung to a hillside where small miombo trees provided spotty shade. Mary had to use her walking stick to get up the steep slope. She asked us not to fuss or follow her so that she could take her time. Julia discreetly kept her eye on her grandmother.

We all assembled at the first rock shelter. We stared at a delicate red painting of three figures with headdresses, painted as though lying on their sides with a pole across them.

"We found a very similar trio not far away, at Kolo 1," Mary explained. "They look very similar, painted the same way, with the same headdresses, but upright, not lying down."

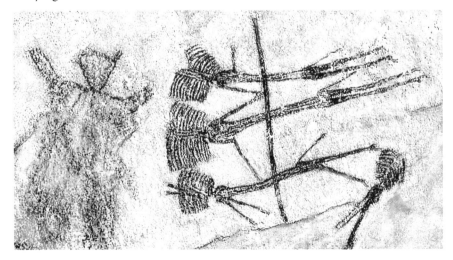

'Three heroes' rock painting, Kolo 2

"What do you make of them?" I asked her.

"It's possible they represent mythical beings, or perhaps the capture of important people."

Turning to the three Hadza men peering at the paintings, I asked the same question in Swahili.

"What do you see in this picture?" I asked. They conferred, looking carefully at the prone figures.

Gudo gave the joint report. "We see this as three champions wearing zebra headdresses and lying with a ladder pole. Maybe they are lying down to rest. Sagiro thinks they fell while trying to use that pole as a ladder. Kampala thinks they used the pole to climb up on top of the rock. Up there, they would be safe from their enemies."

'Abduction' rock painting, Kolo 1

Mary's eyebrows climbed playfully above her spectacles as I translated. Her comment, "Interesting."

The next site was Kolo 1 with a famous scene of "abduction." This one seemed straightforward—two men pulling the arms of a woman, one man with a big erection. The possible meaning was unmistakable. David and Julia tape-recorded the ribald comments we all made, none of which helped us understand the picture any better.

The Hadza agreed with most of Mary's identifications of the animals. That seemed a promising trend until we realized how suggestible the Hadza were. If Mary called a figure a kudu, the Hadza were likely to see a kudu as well. We all tried harder to let them decide for themselves before talking about interpretations. It took patience to carefully look at the faded and overlapping designs.

I was glad that the Hadza could say, "I don't know what that is," about some paintings. They just as often had bright ideas about others. For instance, Mary had noted a mystery figure painted at several different sites. She thought that the figure

'Candelabra' rock painting

represented *Euphorbia candelabra*, a cactus-like tree with many upright arms. The Hadza interpretations ranged from a pit with stakes to an animal hide split into strips. They explained that such leather could be carried on a stick as a camouflage to hunt animals. It could also work to hide from enemies or could even be seen as a dance outfit. They definitely did not view the design as a plant, let alone a candelabra tree. Why would anyone paint a common tree? The Hadza tended to find animals or stories in the designs.

From the third site, we straggled back down to the two cars parked at Mary's old campsite. The wind had panted its way west, leaving us in the sweaty arms of midday heat. Hot and tired, we got into the oven-like cars and went back to our breezy camp for lunch.

In the afternoon, Mary and I sat in the shade of the mess tent and wrote down our confusing notes. David and Julia tried to transcribe the tape recordings. That took hours because they had to translate from Swahili into English and ignore the Hadzane chatter, too. Also, the voices often overlapped, making it hard to understand what was said. We agreed to give up trying to record the conversations. Instead, Margaret and I appointed ourselves as note-takers. We reckoned that our two different versions might get us closer to the reality of the different ideas expressed.

Nigel and some of the camp crew went off to check out the road to Cheke, where we wanted to go the next day. Juma insisted that the road was quite impassable, but that was just a challenge to fearless Nigel. He returned late in the afternoon, saying, "We had to shove some rocks, but it's good enough for us to reach the site tomorrow."

Juma looked pleased at the news. He said with his typical politeness, "The chairman of the Kolo village might join us, if it is acceptable." Indeed, we found it more than acceptable. I liked the idea because it was politic to involve a local politician. Juma revealed another big advantage to having the chairman along: he

Rock shelter at Cheke

just happened to be a local Sandawe man. His ancestors had lived in the Kolo area from times forgotten. Maybe the chairman could share knowledge about his tribe's history and give us a new perspective on the rock art, too.

That night the moon shone even brighter, and the campsite felt like home already. Mary had no complaints about the cold morning as we assembled by a merry fire to drink coffee and tea. After breakfast, Juma and Pius arrived with the chairman, Boniface. He seemed a friendly middle-aged man who expressed interest in the venture.

Down the rift road again we headed to Cheke, an impressive site. We walked up a rocky, dry waterway to a vast rock overhang. The ledge underneath faced into a steep canyon brightened by colorful miombo trees. The largest rock surface swarmed with paintings.

We all agreed to systematically search the site from left to right. That proved as easy as getting frogs to fly. The three Hadza jumped here and there literally and imaginatively as they examined the largest and boldest of the paintings. Boniface and the guides joined in with their views as well.

I stood bemused while people talked, whispered, chatted and interrupted one another. The words poured over us, distracting and confusing in the extreme. Margaret and I struggled to keep up. Fragments about headdresses, skin coverings, skirts, tails, and weapons made it into our notes. The "euphorbia" shapes were there again—this time interpreted as flywhisks! Imaginations blossomed all around.

The most dramatic item on the rock face was a huge yellow figure overlaid with many other paintings. The Hadza had a name for this monster, *Duduklea* (du-du-clay-ah). Sagiro seemed especially impressed. He told us, "The duduklea ate people and other animals." Gudo added, "That throat pouch is where it stored its prey."

Boniface confirmed that the Sandawe also had this myth of the big-pouched monster. Mary had put the figure down as an eland, an antelope with a pendulous dewlap. We felt good to think we now had a different notion of the meaning of the painting. While checking other figures against the illustrations in Mary's book, Kampala excitedly spied a little tracing of what Mary had called hyenas. He stabbed his finger at the picture, exclaiming, "Wild dogs!" We all bent to look at the image in the book and agreed. Even Mary agreed. She had never been adamant about her interpretations of rock art.

After what seemed to me hours of ideas bonging around my head and banging onto my notebook, someone suggested lunch. With relief, we descended the rocky gorge and had our picnic on flat ground, sitting on mats. Slices of whole wheat bread served as plates for our pick of Mary's special Kenyan sausages and cheeses. Gibb's Farm produce added freshness to the meal: tomatoes, carrots, sweet peppers, and lettuce. I contributed my homemade peanut butter, a tried and tested food I knew the Hadza would eat. They sniffed at the pungent sausage and cheese with disgust, then spread the nutty brown peanut goo on their bread, adding honey and slices of banana from the ripening stems we'd bought on the trip.

All of us stretched out on the mats and relaxed, looking up at the rock faces and the feathery-leaved miombo vegetation on the rift wall. I hoped the others felt as contented as I. Belly full after a delicious meal, resting among friends, with adventures behind and more to come. I smiled, stretched out, and fell asleep.

Going back to camp, we stopped in the village at the base of the rift wall to barter for fruit, mats, and gourds. The gourds were huge, the size of platters and punch bowls, a specialty grown by the local Irangi people. The Irangi tribe spoke a Bantu language. They'd migrated inland from the coast a few hundred years ago during the time of the slave trade. They were one of the many Bantu tribes that formed a grouping our Japanese anthropologist friend Wazaki called the *Waswahili*. Most Irangi were Muslims, having brought the coastal religion with them.

We strolled through the village, mixing with the Irangi women who sat in front of baskets piled high with guavas, oranges, tomatoes, and peppers. Although Muslims, they were not clothed in the black *buibui* garments of their coastal relatives. Instead, they wore a colorful mix of kangas.

Back at camp, the Hadza men, the Sandawe chairman, and the guides chatted amicably. I overheard them talking about the way to make the ancient paints out of resins, ochre, charcoal and ash, while the rest of us sorted out our notes. Evening crept upon us, and Nigel took the locals back to Kolo village. I was relieved that all was going well but felt worn out. Trying to keep people moving and focused, guiding them from one picture and rock face to the next, took all the energy I had.

Kampala was especially hard to contain because his lively imagination kept generating bold ideas, pre-empting other people's interpretations. Mary was the opposite; she stood back quietly and watched with an expression of mixed amusement and skepticism. The rest of our team explored around, then came hopping back to distract or interrupt, and generally get in the way. Hard work for me, the frog herder.

I slept hard and woke weary. This day we wanted to visit the Kisese rock paintings. In addition to art at the rock shelter it had evidence of human occupation dated to more than 40,000 years ago. The site was far along the base of the rift wall. Always the impatient one, I set off for a brisk walk before the others got started. I took a shortcut along a footpath, aiming for the main road down the escarpment. Striding along enjoying the early morning air, I noticed a man following me. I didn't pay him much attention until he passed me, slowed, and looked back at me over his shoulder.

I decided to leave the path and head straight to the road. But soon I found myself in a thorny thicket at the edge of the rift wall. I realized that I'd better cut back to the trail, or I'd miss the cars. The man who had passed me stopped and watched me. When I reached the path, he turned to follow me again. That made me nervous. I sped up, but the man kept up close behind me. We intercepted the main road just as David came driving by. The man flagged him down.

"Greetings, Mzee," said the man in English. He pointed to me and added, "Please give this woman a lift," he said. "She does not seem to know the way." The remark amused everyone, even me. "Thank you, Bwana. I'll take care of her," David laughingly promised.

On the way along the valley floor, we passed through the Irangi village again. Market day seemed to be in full bloom, so we alighted to sample the wares among the colorful throng. We bought cooked sweet potatoes, ripe tomatoes, sugar cane, and a sack of white guavas. Most of my companions went shopping for pots, wooden utensils, baskets, and mats. I found a woman who sold me seeds for growing the giant gourds.

The Hadza and I munched on roasted sweet potatoes as we proceeded to the Kisese site.

We were disappointed to find that the paintings there were very difficult to see, faded, and overlaid with graffiti. We tried to make out the iconic flute players that Mary had put in her book. The Hadza looked intently at the figures.

"These are not playing instruments," Gudo announced. "They are smoking pipes. Perhaps they…"

Kampala interrupted, "I don't agree; maybe they are blowing darts."

Sagiro muttered, "No, they are playing the filimbi, the little flute."

We peered and squinted at the vague paintings, tilting our heads.

"My head hurts!" complained Kampala! Putting his hands over his ears he plopped down under a spindly tree. Nigel and Julia wandered off and soon came back to tell us they'd discovered a freshly deserted leopard's lair up the wash. Most of the rest of our team went for a look. Mary and I stayed staring at the vague deteriorating figures and designs. We worked our way along the rock face, picking out images until we reached a picture of a recognizable giraffe and rhino. I suggested we stop there. I felt as smudged as the old paintings.

The rest of our explorers returned from the leopard's lair. While we rested in the shade, Juma asked us the essential question: "Why do you think ancient people painted the rocks?"

We'd heard many interpretations of Kolo rock art and ancient pictographs in general. It was fun to gather fresh ideas as everyone tossed out suggestions:

"For fun?"

"They were bored in the long evenings and needed something to do."

"Personal identification, to say, 'I was here.'"

"Family recognition, to say, 'We were here.'"

"Telling stories or recording events."

"Feeling creative and needing an outlet."

"It was a religious or magical exercise to bring success in hunting..."

To us, the magical and religious interpretations were suspicious. Rock art was all too often said to be made by shamans high on hallucinogenic drugs. It might be part of the story but neither the dope-smoking Hadza nor our Western, alcohol-drinking selves found that idea easy to believe. All the other reasons could be possible too.

Michael Jackson kanga

"I'm *hungry*!" Kampala's loud interjection sidetracked us, and we laughed. Indeed, we were all hungry, so we went to the cars and spread out our picnic on the mats. Kampala sprawled cheerfully, slurping on a ripe mango, dribbling juice onto the soil. Boniface and Gudo wandered off to dig up some medicinal roots while the others ate and relaxed.

Mary and I talked about going to visit yet another site, Kwa Mtea.

With renewed energy, we went. The rock overhang was up a slope of the rift wall. We couldn't make out many of the little animal drawings. By now, we were all too tired to do the painstaking picking through the faded and overlain images. So, we put our backs to the paintings and looked in the opposite direction. It was a splendid view—open grass plains rolling away into the afternoon haze, eastwards across the floor of the Great Rift Valley. A lovely brachystegia tree added a grace point, bending out from the steep rift wall beside us, its leaves fluttering like a flirting girl's eyelashes.

Groaning a little, we got up and made our way down the slope back to the vehicles. Chairman Boniface slipped on a rock and twisted his ankle. He stayed in the car when the rest of us alighted at the village marketplace. While my companions wandered, I searched among the stacks of kangas, those multipurpose cloths I loved. I found a winner—a yellow and purple design of pop star Michael Jackson, greasy hair in the eyes, musical instruments around the border. The kanga was a real modern art treasure, probably the ugliest design I'd ever seen.

The next morning, we headed in a different direction, staying on top of the rift escarpment. We drove a long way west into the Bubu valley, winding along through farmland. The drive gave me enough time to tease Kampala into making up a song in Hadza about the safari. He did, and with Gudo and Sagiro singing, it sounded terrific. The rest of us even caught on to the simple repetitive phrases. Here is the song and the English translation by Bonnie Sands:

Strophanthus eminii - flowers & seed pod

CHAPTER 11: KOLO ROCK ART SAFARI

Mary Leakey at Bubu River rock paintings

Rock Painting Safari Song (Mary Leakey Song)

Likiko	*Leakey,*
musiyobakwa	*she is troubling us*
shauri ya koeta	*because of them,*
Hadzabe kenebe	*the Hadza of old*
hukwa, maha'a	*Get up! Let's go!*
'isawabi'i	*The caves!*
|iyetabita	*We'll see them,*
Hadzabe kenebe	*the Hadza of old*

Singing this song over and over as we rolled along, Gudo suddenly shouted, "Simama!" We screeched to a stop. He and Kampala got out and rushed over to a bush. David and I got out to investigate the odd plant. The cream petals of the flowers were long and twisted like ribbons. It had paired horn-like seedpods clad in nappy grey fuzz. I wondered how it had attracted Gudo's eye. The plant didn't grow anywhere that we knew of in the Lake Eyasi basin, yet the aficionados of toxins had recognized it.

"Very good poison," Gudo told us. We later learned its scientific name: *Strophanthus eminii*. We duly collected the seedpods and portions. We carried on, the Hadza click-clacking to one another about who got what of the poisonous plant.

The Bubu paintings were a delight, big red kudus, and other animals boldly painted on smooth rocks. Everyone could agree on what they represented because there wasn't so much over-painting or graffiti. We got back to Kolo camp late in the afternoon, the nearly full moon high in the sky.

By the evening campfire, we discussed what we'd learned about the rock paintings. We concurred that many of the pictures could have different interpretations. Also, their dating and styles could be studied and argued about endlessly. Our main conclusion: the painters must have been deeply impressed by large wild animals and other humans, friends or foes, not so much by plants. We also agreed that the paintings were in real danger. Would they survive weathering and vandalism long enough to be studied thoroughly?

'Wild dogs' rock painting at Musia 2

The Hadza declared that enough was enough. Those paintings were long ago; the present was more important. They wanted to head home. So did the others. Our group started to split up. Mary, Julia, Margaret, and Per packed up to head back to Karatu. David wanted to leave with the three Hadza to Mangola. Only Nigel and I wanted a chance to do more exploring. In the morning, almost all our group packed up and left.

The camp crew started to dismantle the tents as Nigel, Guide Juma, and I set out to search for the Pahi rock paintings. We drove down the rocky road to the valley bottom, passing through the village. Two blooming citrus trees by a dripping water tank tempted us to park nearby. After a long walk across empty fields, we reached the disappointing site. The painting of hunters, so evident in Mary's book, was covered over with graffiti—names, marks, and signs. A buffalo and another animal were chalked around with white ash.

Juma looked embarrassed at the damage. "The village boys come up here to hunt. They camp and cook in this shelter and ruin the paintings. We go to the schools once a month and give talks, but it doesn't seem to help much. No one understands that the paintings are hundreds, even thousands of years old. Our heritage. A lot of local people still deface the rock art."

Dejected, we walked on. Our final destination was a rock outcrop lying like a beached whale far off in the sea-like mirages of heat haze. We took paths through the dust devils playing over the fields, past dilapidated huts, and empty maize plots. Finally, we came to the overhang. We smiled, relieved, as we looked at the paintings of what Mary had labeled hyenas and the Hadza identified as wild dogs. "Wild dogs," said Nigel. "Definitely," said I. "Yes," affirmed Juma.

We plodded back to the car, sticky with sweat. Our car-guard cut tasty ripe tomatoes for us while we splashed water on our faces. We drank our fill from the tank, inhaling the intoxicating smell of cool water on hot soil. The sweet fragrance of the citrus blossoms mixed with the sharp scent of tomato leaves, making me smile with pleasure. Completing the sensory picture, a Muslim chant floated out from the house nearby. I looked up at the sky, shimmering in hazy golden light—another magical moment to etch in my memory.

We gave Juma a lift home, tipped and thanked our guides. Luckily Boniface, the Sukuma chairman, was there too so we could ask about his ankle and promise to send him pictures. He thanked us for taking him along. "You have renewed my interest in our heritage and this fascinating rock art."

Getting back to camp, we found the hard-working crew taking down the last of the tents. They packed the equipment into their Land Rover, leaving out camp beds for Nigel and me. They would head home while we stayed for one more precious night.

Nigel and I ate leftover food and watched the full moon come up. I felt the warmth of the fire, smelled the wood smoke, and hummed the song Kampala had made up. I could easily imagine the retreating voices of the Hadza men singing their way home with David.

As I settled onto my camp cot in the open, a thought tiptoed into my darkening mind. We'd been too intent on finding meaning in the visual images on the rocks. Vision is only one, even if the dominant one, of our senses. All our senses were strands weaving the meanings for the paintings—the feel of fire, breezes and bodies, the smell of smoke, tastes of foods, sounds of songs, and the stories told around the fire on nights like this.

Looking up at the full moon, I wished I had a stick of ochre and a piece of charcoal. I'd go to a big rock nearby where I could paint that enigmatic plantlike figure with its many arms reaching up to the sky full of our ancestors.

Rock painting of flute player, Pahe 27

CHAPTER 12
KICHAA
DO-GOODERS AND KOOKY TOURISTS; WHAT THE HADZA SUFFERED: 1999

"Kichaa" emerges from a Hadza hut

Some visitors to Mangola were fun, engaging, challenging, or just plain baffling. Kichaa, the Swahili for a crazy person, was a puzzle that had lost some of its pieces.

We didn't know who he was when we first saw him at Chemchem Springs. We saw only a skinny, pale European man washing clothes next to his brown-skinned Tanzanian driver who was throwing water on a big truck. Washing vehicles and clothes there was an activity all of us residents discouraged. It polluted the springs, right at their source. But no village officials or campsite caretakers were around to protest.

"Should we say anything?" I asked David as we walked by on our way to Gorofani village. "No, better not; remember we're trying to stay out of conflicts around the springs and campsites." Trying to help had got us in trouble too often. So instead of commenting, we called out "Hamjambo" in Swahili, greeting the two strangers. No response. We then tried the English version, "Hello." The men ignored us. We waited a while, watching them.

The lanky man stood up, mumbled something, then returned to rinsing his shirt. We didn't understand what he'd said. The truck driver offered us some information in Swahili. "This mzungu is from Germany. He has come to research the Hadza."

We were immediately curious because researchers and their subjects were always of interest to us. And a German researcher of the Hadza who didn't know local customs? Suspicion followed curiosity. "Look at the truck," said David, "it's HH's."

HH stands for Hadza Helper. He had a real name, but we used HH as a nickname. HH was a European, one of the most stubborn of the self-appointed saviors who came to Mangola to "help" the Hadza. In the 1970s, he'd become fascinated with these traditional foragers. The story goes that he lived with the Hadza on and off, coming for short periods over several years. During that time, he married a Bantu wife. HH was a champion of the "primitive" peoples of the world; the Hadza tribe fit perfectly into his passion.

HH was part of an indigenous rights group called Friends of People Close to Nature (FPCN). The organization FPCN claimed it supported indigenous people. HH saw himself as helping "wild people" resist enforced assimilation into Western values and practices. He had strong opinions about what help meant. His mission locally seemed to be to save the Hadza from modern civilization.

Wild of hair and beard, HH seemed less a fanatic than an overly serious academic when I first met him. Even so, he'd earned a local reputation as a troublemaker, remaining elusive, going around the Eyasi area to find Hadza and stir them to fight against authorities. Sometimes he camped at Chemchem Springs and always refused to pay any village fees. Our encounters with HH usually took place at the springs, his big truck parked under the fig trees. The truck was an impressive beast, all kitted out as a camper van, with beds and cupboards. At first, when HH seemed friendly, he proudly showed us his truck. But gradually our relationship soured when he started accusing us of interfering with the Hadza.

We failed to understand how we interfered in an adverse way. However, we did realize that just by living in the Hadza realm we affected them, interacting daily, trying to do what they asked of us, trading with them, building friendships, and helping with some heath care and education. HH helped us examine our own position relative to "helping" the Hadza.

Now, with this alleged researcher arriving in HH's truck, David's and my antennae vibrated. Legitimate researchers often camped at Chemchem with the blessing of the village officers. We wondered if this German fellow had signed the village register. We had heard nothing about him from our three foster boys who passed through the Chemchem oasis daily. I relied on them to report on anything happening at the springs. Jumoda also came to tell us about visitors because he saw himself as the primary local guide and tourism director.

CHAPTER 12: KICHAA

We stood watching the odd-looking fellow, trying to figure out how to ask about his "research." He glanced up at us and started filling a bucket, apparently agitated by our presence. No German words came to mind, so I tossed out a question in English, trying my best to sound friendly. "It will be interesting to hear of your research here. Have you some published papers we might read?"

"*Nein*," he said, not looking up. Hmmm, I thought, he understands English. David asked the crucial question, "Do you have a research permit?" No answer. He turned his back to us and went behind the truck, making it clear he didn't want to talk to us. We departed

Soon the aims of this man became clear. Our three boys, various Hadza, Jumoda, and village officers came to tell us about the "Kichaa" who had moved in with the Hadza. Kichaa acted strangely, pestering the Hadza to take him hunting, following the women around, never taking notes that might imply doing research.

We concluded the Kichaa had come experience to life with the "wild" Hadza people. Compared to other tribes in Mangola, this small group of surviving hunter-gatherers attracted the most attention. A steady stream of visitors came to see, study, film, and attempt to influence them. The list included tourists, government officials, reporters, local and foreign researchers, foreign hunters, health workers, and missionaries from near and far.

The Hadza tribe had fewer than 2000 members. They were spread over a vast area and unfamiliar with group action. They had difficulty defending themselves against the people trying to make money or careers from them. All the "helpers" disturbed and confounded them as well as offering amusement, advice, dollops of food, money, clothes, Bibles, dope to smoke, and other "gifts."

In the 1960s, Julius Nyerere, Tanzania's first president, called the Hadza to come out of the bush. Most Hadza declined the invitation. They didn't want to become farmers so hid from outsiders. They feared conscription, taxes, and having to live in one of Nyerere's socialist-style ujamaa villages.

The free-ranging Hadza were seen as primitive, an embarrassment to politicians. The government made several efforts to get them to change. Each time these people were pushed into villages, they got diseases and distressed. Many headed back to the bush. With the bushlands intact, some could survive there. But gradually their range dwindled. A mix of many different tribes came onto traditional Hadza land looking for grazing, farmland, work, and gemstones.

Besides the local conflicts over land and water, the Hadza suffered the presence of what we called the do-gooders. These people had some characteristics in common—a personal vision of utopia, usually involving religious conversion. Most do-gooders seemed bold and self-righteous, lacking sensitivity for the culture and history of the Hadza and other tribes in Mangola.

A widespread assumption held by people like HH, and his friend the Kichaa, was that the hunter-gatherers did not want to change. In the Hadza case, I tested the assumption. I asked several Hadza friends to choose between:

Hadza people under pressure from their helpers?

1) Bags of grain from donors; cash and t-shirts from tourists; housing, hoes, and seeds from the government.

2) Their homeland protected from outsiders so they could continue to live where and how they wanted; to gather fruits, tubers, honey, as well as hunt wild animals.

After pondering, their answer was usually, "Both," or "All of that." Most knew they had to find some balance and compromise or disappear as a tribe.

The stranger who we came to call Kichaa was neither a do-gooder, nor a researcher. He turned out to be a bizarre type of back-to-nature tourist—a Hadza groupie. He stayed at a Hadza camp on the little scarp above the Chemchem Springs.

One day I passed through that camp. I paused to chat with Silsi, a Hadza woman I knew. I caught a glimpse of Kichaa and stared. He was rolling out a mat wearing only a loincloth. He pretended not to notice me, and I returned the favor. I questioned Silsi about him and gathered that the Hadza didn't know what to make of him. In their amused way, they tolerated his presence. Silsi laughed when looking at Kichaa. I wondered if they liked him, found him good entertainment, or were too polite to tell him to go away.

Others were not so forbearing. Village officials told Kichaa to sign in at the village office. They said he should not camp with the Hadza, but he would be allowed to use the public campsite at Chemchem. They questioned Kichaa about his alleged research. Kichaa seemed unconcerned about not having a valid research permit from the Tanzanian authorities.

CHAPTER 12: KICHAA

Kichaa had a habit of staring at a speaker with a blank expression or avoiding eye contact. No one could decide if he was stupid or pretended not to understand Swahili or English. In any case, he simply ignored anyone's suggestions about camping, permits or permissions.

Kichaa stayed several days at the Hadza camp. According to those who came to our compound regularly, his "research" included bathing with the women and sleeping inside huts with whoever would allow him. He even went about naked among them.

But one night, Kichaa tried to get in between a man and his wife in their hut. The couple pushed him out. Ever polite, they told Kichaa he could not have sex with a married woman. They told him to go sleep with a well-known prostitute in the village. Grumbling in German, and rattling the sticks of the couple's hut, Kichaa went to sleep elsewhere.

Eventually, all the Hadza at the cliff camp became weary of the crazy man. "We tried telling Kichaa to leave," said Magandula, the elder of the Hadza camp. "He pretends not to hear us."

The Hadza became upset enough to go to the village leaders. For the Hadza to ask officials for help was an act of desperation. Such unusual behavior told the village leaders something was very wrong. Cleverly, our village chairman went to the local German couple. Johannes and Lena Kleppe were current owners of Mangola Plantation, trying to survive by growing vegetables and fruits for the tourist lodges. They would be able to speak German to Kichaa.

Johannes was away when the village men arrived, so the officials asked Lena to go over to the Hadza camp and talk to Kichaa. "Please persuade him to leave," they told her. "He is causing trouble. And he has no permits. The Hadza want him to go away." Several of the village leaders accompanied her to the camp.

Lena, our modern Brunhilda, told me what happened. "When I arrived, I saw this fellow wearing a kanga wrapped around his middle. He had a Hadza child tied to his back! I tried to talk to him in German. I asked him, 'What are you doing here?' He didn't answer; just ignored me. I told him the Hadza did not want him to stay with them. He just shrugged as though it was none of my business.

"I tried to tell him to leave before there was trouble. He finally raised his head and looked at me." She demonstrated the look he'd given her, a blank, wide-eyed intense stare. I could imagine the scene—the skinny Kichaa facing the tall, blond woman, the village officials standing back, the Hadza all around, listening.

Lena went on with her story.

"He stood there and tried to look down his nose at me. He put his hands on his hips and told me in German, 'No, I am not leaving. I have not finished my research.' I was astounded. I had to look for help from the villagers who stood around doing nothing. What could we do?"

Lena and the rest of the Deportation Committee conferred in Swahili. Lena then translated their conclusion into German and confronted Kichaa. "If you have not left by tomorrow" she told him with all the sternness she could muster, "the village militia will come here and move you out by force." Knowing our village vigilantes, that was a serious threat. Kichaa ignored her.

That same night Johannes Kleppe came home. When he heard this news, he was delighted, eager for a fray. Here was his chance to be a hero, protecting the Hadza, his wife, and two children while enforcing the mandates of Gorofani village, too. It was a role Johannes was born to play.

Morning came. Wearing his necklace of porcupine quills and a fearsome knife on his belt, Johannes was ready for the showdown with Kichaa. Armed with determination, anger, and righteousness on their side, the militia, Johannes, Lena, and village leaders assembled at the Hadza camp on the cliff. Kichaa was there, ready to fight back. Immediately, an argument erupted in German.

Kichaa said again, "I have not finished my research. Here is a free country. I can do as I like." Then he added what he intended as an irrefutable statement. "The Hadza chief at this camp has invited me to stay."

Johannes instantly retaliated. "You don't know much about the Hadza. They don't have chiefs!" That was true indeed, and everybody looked around to see if one of the Hadza would step forth and admit they had pretended to be a chief. Had anyone invited this crazy man to stay? No one made a move.

Kichaa frowned, then pointed to a Hadza man and a woman who had been standing well away from the vigilante committee. "Those people there, they have invited me to stay." The Hadza couple grasped enough of the argument to speak. They looked at Kichaa and said in firm Swahili, "Kwa heri"—a definite goodbye. Kichaa pouted and stood fast; he was not about to move.

Johannes would not back down either; after all, he was to be the hero of this confrontation. He looked at Kichaa sternly and told him in German, then Swahili, "You must leave immediately. We will all wait here to ensure you go. Then we will follow you out of Mangola."

Kichaa seemed to realize he had no support, so he reluctantly started to pack up. He didn't have much; presumably, what he did have was still in the truck that the driver usually parked over in the village. The driver would be necessary to take Kichaa out of town. Johannes, a giggling group of Hadza and all the onlookers, followed Kichaa into the village. They found the lorry driver having a drink with his buddies. Everyone cheered when Kichaa got into the truck that took off up the Horrid Road away from Mangola.

But it was not the end of Kichaa. Halfway up the road, he offered the driver money to return to Mangola. Oddly, the driver refused. Either he realized it wasn't a good idea, or he wanted to get to Arusha with the load of onions he'd bought in Mangola.

However, the driver did come back, and yes, he brought Kichaa. This time, Kichaa went to the village office. He told them he was going to camp with the Hadza again. He didn't mention doing any research; this time, Kichaa played the

CHAPTER 12: KICHAA

do-gooder role. He said he brought food to give to the "needy people." Always willing to forgive outlandish behavior, the officials didn't stop him. Off he went. The lorry driver parked the truck on the cliff above the Chemchem Springs near the Hadza camp and promptly disappeared to the village.

A few days passed quietly because no one seemed to realize Kichaa was back at the camp. Then came the dramatic climax of this farce. A Japanese film crew arrived. They had come to film the Wild People of Mangola. Two cars drove onto the little plateau and pulled to a stop beside the camp. The Japanese piled out of their vehicles to set up their equipment. Did they have permits, and had they checked in with anyone? No one seemed to know. But there they were, cars, cameras, directors and directed, in a milling confusion.

Japanese voices got louder as they gesticulated at the truck parked by the camp. They made it clear that they wanted the white man and his truck out of the scene. Likewise, they didn't want the skinny Kichaa parading around in their film of the wild bushmen.

Politely, one of the Hadza men, tried to persuade Kichaa to move to the periphery of the camp. Kichaa got angry. He shouted and waved his fist at the Japanese and at the Hadza, too. Something must have snapped in his mind because he rushed at a Hadza man, pushed him, and made a grab at his handful of arrows. The Hadza man hung on. As they struggled, one of the poisoned points scraped the Hadza's hand. He let out a yelp.

Fellow Hadza and members of the Japanese film crew rushed to help. Wounds from poisoned arrowheads could be life-threatening. The Hadza tried sucking out the poison, they wiped away the blood, the Japanese offered medicines. Kichaa stood aside, ignored in the melee. An observant witness realized what could happen and ran to the village to fetch the truck driver. The driver rushed back, thrust a wad of banknotes at the wounded Hadza man, shoved Kichaa in the truck, and fled.

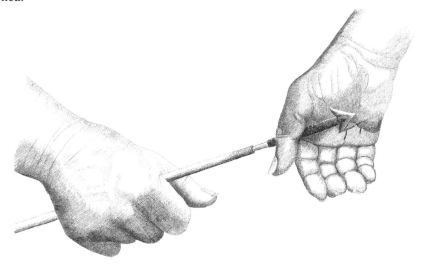

The Hadza man survived. Kichaa did not return. Even HH faded from the scene. Feeling relieved, the rest of us just hoped that the Hadza tribe could continue to survive the stings and arrows of outrageous fortune from those who come to "help" them.

Hadza arrow-heads, forged from large steel nails. The left point is used for large animals, its base wrapped with poison gum and loosely fitted into a wooden shaft. The center point is used for fish. The leaf-like point is used without poison for small animals like dikdik - the blade causing enough damage to maim or kill the prey.

CHAPTER 13
SCHOOL GIRLS
LESSONS IN LEARNING: TRAUMAS, TEARS, AND TESTS: 1996-2002

Schoolgirls Rega, Asha, Suzanne and Mariamu, dressed up in kangas

I had a quest. I needed to find schoolgirls. Among all the tribes and cultures in the huge Lake Eyasi Basin, I was looking for Hadza girls, ones who could make use of a rare educational opportunity. My best hope of success was the school in Endamagha village at the north end of Lake Eyasi. There, the Tanzanian government had built a primary school for the dwindling numbers of Hadza foragers. Officials rounded up the reluctant children from their remote camps in the wild hills and valleys and trucked them to the boarding school.

Driving to Endamagha was an obstacle course: ditches, washes, and ravines; cattle and goats with their grim-looking Datoga herders; thorn branches, rocks, dogs, and children in the road. Shaken from wrenching the Land Rover to and fro, my dust cloud and I made it to the schoolyard. Block buildings squatted in a clearing dotted with scattered thorn trees.

The school was still in session, the murmur of children's voices audible through open windows. The door to the headmaster's office was open too, letting the hot breath of the day enter with me. A stout, unsmiling man sat at a cluttered desk. He nodded his head as I introduced myself and told me to sign the visitor book. I did so, slowly, while mentally composing sentences in my inadequate Swahili. I readied myself to explain why I'd come.

The story behind my quest started weeks before when David and I visited one of our favorite places in Tanzania, the Tarangire National Park, a tented camp among monumental baobab trees. It overlooked herds of elephants in a sandy river bottom, giraffes browsing tall fever trees, and baboons barking from rock outcrops.

One evening at dinner, we sat with the Simonsons, who were also staying at the lodge. Dave Simonson was a large, white-bearded American who could probably toss a bison over his shoulder in his home state of Minnesota. His wife Eunie was also a big person, a nurse who looked as though she could cheerfully carry armfuls of little children. They fit the image of larger-than-life people with hearts and reputations to match. They'd spent many years as Lutheran missionaries in Tanzania, a dynamic pair with a legion of devoted followers.

I'd come to know Dave while helping to set up a conservation education project for Tanzania. Part of my team were two men, one from the National Parks and the other from the Wildlife Division. Early on, we decided that we needed to broaden our support, so I visited various religious groups. In addition to their focus on human welfare, I tried to assess their attitudes towards the Earth and its wildlife. Most local church leaders seemed unconcerned about our planet. Dave was different; he strongly supported environmental education.

At dinner, Dave introduced me to a comely young woman sitting next to him. "This is Jane, our headteacher. She is helping with our new project." They explained that the church was building a secondary school for Maasai girls. Dave believed the Maasai could no longer take their cows anywhere they liked, steal cattle, kill lions, marry off daughters at an early age, and keep women subservient. He was convinced that the Maasai needed educated women. Those women could become community leaders, able to help face modern challenges.

For this big, burly patriarch to be so eager to change a patriarchal society into a more gender-equal one amazed me. Dave told me that while the school was still being built, he and others were trying to coax the Maasai elders into accepting the idea of education for their daughters. Dave was indeed a most devoted coax artist. He'd already established many churches and hospitals in the region; he would get the school built. I was awed by his energy and perseverance. Feeling humble and inadequate partly set me up for the shock to come.

"We have kept four places for you," Dave said with a gleam in his eye. I looked at him, puzzled. He grinned, obviously pleased with my reaction. I turned my surprised face towards Jane. She also smiled, nodded, and said, "Yes, there are four places. We are hoping you will fill those places with girls from the bush."

CHAPTER 13: SCHOOLGIRLS

Both watched me as my thoughts floundered about, trying to think of where four girls might come from. I heard Jane say, "We are keen to have four Hadza girls. But you need to find some who are top-notch students."

"Yes," said Dave, "and they must be willing to live away from home with Maasai girls. We're planning on having about 45 girls each year. We know that most recruits will need several months getting up to speed." He looked at Jane who added, "After some preliminary brushing up, they're going to tackle physics, chemistry, biology, more math, more language, civics, history, geography, religion, and agriculture!"

What a lot to learn! The coursework sounded daunting for anyone and most certainly for students from our understaffed and undersupplied local schools. Well, I thought, getting them through such courses would be the school's problem. First came the task of getting any girls who would undertake such challenges.

Dave's voice cut through my thoughts. "I firmly believe this is what the Maasai need for a better future. And I remember that you, too, have a passion for education, for teaching girls in particular. And you also know and care about the Hadza. So, there you are: four places to fill!"

And here I stood, in front of the headteacher at the remote bush school trying to explain my mission. The gist of what I said in Swahili was that I'd been allowed to choose four top Hadza girls students from his school and give them a chance to go to secondary school.

He slumped in his chair with a frown as he pondered my words. I filled the silence with a request, "Please, could you introduce me to a few of your best students?"

He agreed and left the office. While he was gone, I peeked at the visitor book filled with swirling signatures of politicians and dignitaries. I wondered what they came for: to see the wild-caught children sing or dance, or to be seen seeing them? What did this bush school have to offer? It had not yet produced any students good enough to pass national exams for entry into secondary school. Teachers disliked being at Endamagha and most of the parents hated having their children taken there.

I knew conditions were difficult. I'd raised funds and given our own money and labor to help build a block of toilets. I'd bought mattresses and books for the school. The children lived in overcrowded dormitories and were fed unhealthy, paltry meals. They often fled to look for food outside the school. Discipline was often harsh, and there seemed little interest or sympathy among the staff.

I followed the headteacher outside and waited in the shade, trying to feel more optimistic about the school. He returned across the schoolyard with a girl who looked down at the ground the whole way. She was plump for a Hadza, with very short hair and a rumpled school uniform. Stopping more than an arm's length away, she raised her face and looked at me. As we gave each other a once-over appraisal, I could see the intelligence and suspicion in her brown eyes.

Headteacher introduced her: "This is Mariamu, my best student."

Mariamu surprised me by giving an oddly engaging little curtsey as she extended her right hand, holding it at the elbow with her left. This was a very formal way of greeting; I was impressed.

Mariamu murmured "Shikamoo"—she-ka-moo, the Arabic word meaning *I hold your feet*. I returned the correct response, "Marahaba"—mar-a-ha-ba, also an Arabic word, implying *thank you*.

"Mariamu is finishing her seven years of primary school," said the headteacher. "She hasn't attended all of every year. But she is smart. So smart she's somehow stayed at the head of her class." His lips curled up—was that a smile? I smiled back. Mariamu's face remained expressionless.

She stood aside as the headteacher called over another of the top Hadzabe students, Asha. She shook my hand limply and stepped back. But then she raised her head to look me in the eyes and nodded, saying in a low voice, "Jambo, Mama Simba." I was mildly surprised that she knew me and wasn't afraid of me. Possibly she'd make her way among the Maasai girls, too.

The headteacher brought me back to attention, telling me about the other top Hadza student, a girl named Suzanna. "But she has gone home to the Hadza village in Yaeda Valley. I have another excellent student who is here today. Maybe she, too, would be suitable?" He waved over a statuesque girl I knew was not a Hadza. The headmaster confirmed this by telling me, "This is Rega, she is a Datoga and she gets very good marks."

I looked at the tall beauty looking down on short me with what I hoped was not haughtiness. We pressed our hands together and I knew I'd try to get her in, too. I'd been given four places for Hadza girls, but now I had the chance to add a representative for another neglected tribe. Also, Rega knew the three other Hadza girls. Together they could support one another and maybe hold their own among the Maasai girls. Or so I told myself.

I submitted my choices to headteacher Jane at the secondary school. Exclaiming with delight she told me, "Oh, I do like the idea of including a Datoga as well as the three Hadza girls. I'm eager to meet them all."

When I returned to Endamagha to tell the headteacher that the new school would accept "his" four girls, he seemed proud. "Very good. Yes, very good. This will be a first for our school. Finally, we have students from Endamagha primary school going on to secondary."

His statement reminded me that these girls hadn't yet passed the national exams. Maybe they were only outstanding relative to the slower students in the bush school. I still worried about how the four girls would cope with Maasai girls, traditional antagonists to the Hadza and Datoga. The new secondary school was in Maasai territory and our four would be very far from home.

I wanted to find out more about the four girls. I needed to prepare them for their trek up the slopes of higher education, so I asked them to meet me at our place on the next market day. Mariamu, Asha, and Rega came early. The missing

Suzanna arrived later, separately. She was a serious-looking young woman who set herself apart from the others.

I had devised a series of questions in English and Swahili, made up some math problems, and added a few questions about history, geography, and biology. In addition, I asked about their home life, interests, and ideas about their futures.

The girls wrote their names and ages on the top of the quiz papers. They were all older than they looked—18 and 19 years—meaning they'd started school at the age of 10 or 11, not the national required age of 7 or 8. Sadly, none of the girls seemed adept in any of the subjects I quizzed them about. They were especially poor in English, which would be the language of instruction at the secondary school.

I decided to start tutoring them, getting them used to a white-woman teacher in the process. Mariamu, Asha, and Rega stayed with us for short periods and stumbled along, doing well in geography and science subjects, but struggling with English and math. Suzanna seldom came to us but wrote letters and read books on her own.

On a supply trip to Arusha, I stopped by to see Jane at the partially finished campus. Many sturdy buildings were growing up among rows of coffee on what had been a plantation on the southern slopes of Monduli Mountain. There was a large new sign: Maasae Girls Lutheran Secondary School. The name MaaSae was an acronym formed from Maa, the language of the Maasai tribe, with S for Speakers, A for Advanced, and E for Education. I smiled, knowing the name would flummox many who came to the new school.

"Jane, quite frankly, I'm worried how these four girls will fit in. They come from a rural primary school with only basic academic values. They don't know as much as they should."

She tried to reassure me, saying, "Many of the Maasai girls aren't at all proficient either. Our best approach is to start rural students in Pre-Form 1."

"What exactly is that?" I asked.

"Pre-Form 1 is a period during which we teach special classes, assessing their abilities in English and Swahili and their knowledge of history and math. That time allows them to settle in with the other students too."

"Sounds good to me," I said, and mentioned I was doing much the same already.

The four girls got their four places. David and I took them to school, where the teachers and other students welcomed them with real warmth. My skepticism about how they would adjust was replaced by relief when we checked on them in the middle of the term. The four came cheerfully to greet us in their crisp red and white school uniforms. They seemed enthused about their friends, teachers, and classes. We awaited further reports.

Some news arrived in the form of a letter from headmistress Jane, leading to unexpected developments.

"Greetings, Mama Simba. Thank you again for bringing Mariamu, Asha, Rega, and Suzanna to Maasae Secondary School. They all did well in Pre-Form 1. They will now start regular Form 1 in January. I am trying to think of ways to help them use their native languages more. Singing is one way. Asha taught the gals a song in Hadza with a Swahili translation. They sang it on parents' day. It's a beautiful song."

I was happy to get this news but not so happy with the next part of her letter.

"We have a bit of a problem. Mariamu is more than six months pregnant. She said she was pregnant before she finished standard seven. Mariamu asked me to write to you, as you know her and her mother."

I wasn't too surprised, because I'd noticed Mariamu was somewhat pudgy when we first met. But in the excitement of getting the girls into school, I'd set the suspicion aside. A deeper concern was that Tanzanian law prohibited pregnant girls from continuing school. Then I remembered that Mariamu had not yet officially started secondary school, just done Pre-Form 1. Hopefully, she could not be expelled.

Jane's letter continued: "Mariamu is not the first of our new students to come to school pregnant. She is the seventh. Each case is different."

Mariamu's case was now mine too. I read the ending of Jane's letter more than once. "Mariamu is welcome to stay here. She is showing. The girls are accustomed to being around pregnant women. Mariamu said she hopes her mother might take care of the child. She wants to continue school after the baby is born. Is there any way her mother could come here when the baby is due? The alternative is to send Mariamu home. She is extremely bright and wants to study so I will do all I can here. Let me know what we should do."

As soon as I could, I stopped by the school to talk to Mariamu. Her pregnancy was very visible, emphasized when she put her hands over the mound of her belly. She looked down when I asked about the father of her unborn child. "He was my teacher. He is gone now. They transferred him."

"Do you want me to try to find him?" I asked.

"No!" she said emphatically, her eyes welling with tears.

I forged ahead with my next question. "I know you want to stay on but tell me what you want to do with the baby. You can't keep it here at school. Do you want to give the baby to your mother?"

Mariamu shook her head and clenched her hands. She said in a sad and bitter voice, "My mother is a drunk, she buys *pombe* (booze) when she gets money from tourists. I'm not sure she could take care of my baby, but I don't know anyone else who can."

I paused to think whether or not I'd be willing to take on the newborn. Nope, not possible, I quickly decided. We were doing enough parenting with three abandoned boys and our foster daughter Ruth in nursing school. We didn't want to raise a small child. And I didn't want to take a Hadza baby out of the tribal setting.

"Mariamu, I will look for your mother. I'll ask her to come here and help you during the last part of your pregnancy. Do think about what you want to do with your baby."

Back home, I learned Mariamu's mother was off in a Hadza camp far away. I sent requests for help via my Hadza friend Abeya. The bush telegraph worked, and Mariamu's mom showed up on a market day. We sat together, making efforts to understand one another. I couldn't speak the Hadza language, and her Swahili might have been worse than mine. She wasn't drunk but seemed unable to take in what Mariamu's pregnancy meant.

Most of all, she didn't want to go to her daughter. I worked hard to convince her, promising to take her myself, giving her new clothes, toiletries, pocket money—in short, I bribed her. She finally agreed and we settled on the next market day to meet.

I sighed in relief when she showed up. She sat silently most of the way up the Horrid Road and said little when we stopped in Karatu town for a soda for her and coffee for me. Another couple of hours later we reached the school. Mariamu's mom looked around with a mix of fear and apathy at the modern buildings and ranks of coffee bushes beneath the tall shade trees. The scene must have seemed a world away from the near-desert of Mangola. She showed no delight in the reunion with her daughter. Mariamu came out of her dorm room and guided her mother away.

Alas, Mom stayed with Mariamu just long enough to see the baby home from the hospital. She then took all the remaining money, sold the new clothes, and made her way back to Mangola. There she disappeared into the bush.

Mariamu chose to stay with a missionary family living close to the school. The couple offered to look after her while she nursed the baby. I went by to see her and her little boy as often as I could. I smiled when Mariamu told me, "I've named him Akili after the man in your book."

Akili meant "smart" in Swahili. It was the name we'd given the main character in a book about small-scale farming. We contrasted the smart Akili with a slow-witted companion named Zuzu. At that time, all we could do was hope Mariamu's little Akili would live up to his name. Eventually, Mariamu returned to school and left Akili with the missionary family. I made several attempts to make it easy for the child to return to his mother's homeland to be raised by relatives. But Mariamu did not help. I gave up and Akili stayed with the missionaries.

The following year, I was allowed to submit the names of four more girls. I took more care to check on possible pregnancies. All seemed fine to start, but one soon had to drop out when she, too, got pregnant. I wondered if girls would ever be allowed sex education and means for birth control.

The following year, the system changed. A new woman became headteacher. She seemed less interested in mixing different tribes, and the school became weighted almost entirely to Maasai. The next year, more exacting entrance exams and rules were put in place and I gave up submitting names.

"My" first four girls sang and slid academically through their four years of secondary education. When they came to Mangola, I'd tutor them in English and any other subjects they wanted, including typing lessons on our laptop computer. Since the typing program was in English, it also helped them learn a whole new vocabulary. Asha got married while in school and soon disappeared. The other three stayed with us after their final term. We all waited nervously for their final exam results and were delighted when they passed.

While the second batch of girls was going through school, I pondered what the first survivors would do next. Asha had already made her decision. Suzanne returned to Yaeda Valley to take part in that Hadza community. Mariamu and Rega went back to Mangola with me.

Rega's post-graduation story was not encouraging. She went home to her brother's place where she found it difficult to fit in. She wrote me a letter to explain how unhappy she was with her situation.

"When I arrived there at my home, I greeted people. They just stared, looking at me. They asked, 'When did you come from Arusha?' I told them I was not studying in Arusha but at Monduli. None of them have traveled far from here. They do not know about towns. 'What job do you have now?' they asked me. I told them I don't have job. They asked, 'Do you have a husband?' I told them, no, I want education, not husband.

"They told me that I no longer belonged. They told me I wanted to be in Arusha town and I needed nice things. And I said 'No! Mangola is my home.' I tried to speak with them. They didn't listen. They told me, 'You forget about your tribe. Look at you. You wear clothes of Swahili instead of your tribe.' I tried to tell them education is good, for health and life, and they answered, 'We are living a better life than you who has education.'

The rejection depressed Rega. She and I tried developing a job for her using a mobile education kit we put together. She attempted to teach her young brothers and sisters at her home boma. But she didn't stick with it, finding it boring. No one seemed to want to learn anything from her, and she soon got discouraged and gave up.

Sometimes when Rega came to Mikwajuni, I'd give her magazines and books to take home to read. Even after four years of English, she could not read easily. Her favorite job was helping me to plant baby trees and sew. She told me she wanted to learn to drive and type. I set her more typing tasks and let her use my sewing machine. She was still restless and went to town often.

I could understand why her tribemates teased her. She refused to dress like a typical Datoga woman or marry a local man. She liked to wear fancy clothes and acted like a magazine model. Once I saw her at the monthly market, looking gorgeous in a fashionable outfit; no kangas or cloaks for her. I realized she would never be happy in Mangola. Education had made her different, changed her views about herself and her world. On her initiative, she did go to Arusha town. She borrowed money to take a course in tour guiding, writing us sometimes to tell us she was doing well.

And Mariamu? She matured, deciding that she wanted to be a spokesperson for the Hadza. "I want to learn about different tribes and society. I want to be able to tell people who come from different countries about my culture and societies like the Hadza, Datoga, Iraqw, and Maasai."

While she stayed with us, I also gave her a job as an itinerant teacher. She went out to the Hadza camps and tried to teach children to read and write. The aim was to help prepare the children for school, as well as giving Mariamu some status and income. Here are three extracts from her reports, in her own words:

Report 1: "It was the first day to go to this camp to teach. I was to teach the names of animals—mammals, the carnivores, and other kinds. Yes! It was so fun for them. Then I try to taught them how to write their names. Some of them were good understanding, but some of them were not able to do this. Also taught them how to write AEIOU and letters, some words and alphabet."

Report 2: "This day I was with Elizabeth. She is Gudo's wife. She helps me to give her friends books and show them how to use because I have already taught her. She is good and bright, understanding well than others. In this trip I brought your big map of animals. We tried to call those animals on map in Hadza language. Some names were so hard, such as Tsonkwanako. In English is the name of the giraffe. Many animals were not new for them such as baboon, elephant, kongoni, and zebra. This was so fun for everybody."

Report 3: "This day I was with Salibogo, Kampala, and Adamu. I was with them with all my books. In this day I taught them about the two farmers, Akili and Zuzu. Even though they do not understand the words but just pictures, I showed them there was a person who was clever and person who was stupid. It was so fun because no one wanted to be Zuzu. Everybody decided to be smart like Akili."

Whenever Mariamu mentioned the hero Akili in one of the books she was using to teach in the bush, I thought of her son. He was still being raised by missionaries in town, another Hadza child taken from the tribe. She never mentioned him.

Mariamu's English improved with these written reports, as well as conversationally. We recommended her to colleagues and friends as a research assistant and guide. She did become a spokesperson for the Hadza. She has also attended conferences, guided filmmakers, and participated in various political events. Mariamu is still a link between the many tribes of Mangola and with the wider world, too.

Anna Phillipo from the second batch of students also became a champion for the Hadza. She is now on the board of the Olanakwe project that deals with education and development with Hadza people.

What do we conclude from educating these girls through secondary school? I continue to believe that education is important as evidenced by some of the gals who became significant figures in the wider social scene. However, the type of education on offer needs re-examining. And unless there is support once these students complete school, well-meant efforts can leave them marooned between worlds.

Motivation and aims also make a difference in how modern formal education affects youngsters. Mariamu and our foster daughter, Ruth, the nurse, went through school with clearer aims and lots of fortitude; the others had more difficulty. For all the various students we mentored, we can but hope that they, their families, and communities benefitted in various ways. And I certainly received an important education from each and every one.

Maasae Girls School graduates, 2002. Back: Suzanne , Mericiarana, Pili, Ana. Front: Regarista, Mariamu, Upendo, Jeannette

CHAPTER 14
THE FOOTPRINT MAN
MAKING TRACKS, HOW SCIENTISTS WORK, A FRIEND: 1993 ONWARDS

A chimpanzee's foot. With an opposable big toe, it looks more like our hand than our foot.

Take a look at your feet. Bare feet. Are you impressed, or at least curious about them? So was Charles Musiba, who was tracking the evolution of human feet. We met him in Mangola and called him the Footprint Man. His work and career gave us a much firmer grounding on the continent of Africa, where our ancestors developed their bipedal trait.

Our ancient human-like ancestors were two-footed, upright apes with long arms and grasping hands. More than four million years ago, upright apes with small brains lived all over East Africa. Walking on two feet freed hands to hold fruits, babies, gourds, meat, weapons, and mobile phones (well, those came later). Hands are wonders, but so are feet. They carry our weight around and have evolved particular shapes. Without getting too technical about the foot's anatomy, take a look at your toes.

Around 40,000 years ago, the bones in the big toes of humans started to become smaller. Some anthropologists think that's when people began wearing shoes. The oldest shoes found so far are only about 10,000 years old, but those shrinking toes implied we probably adopted footwear some 30–40,000 years ago. And we have been designing shoes ever since, by the zillions.

This story is about feet, in and out of shoes.

I sat on a stool sorting beans at our communal table under the shady acacia tree at Mikwajuni. The approach of feet wearing sandals made of old tires made me look up. My eyes met those of Jumoda, coming from the river path into our outdoor kitchen area. As a self-appointed manager of Mangola tourism, he often came to our place to complain about campers at the village site. Now Jumoda told me, "The group of campers at Chemchem are not locals, but they refuse to pay the fees."

Not that I could do anything about that. The entire Mangola area was swarming with tourists. Our nearby village had received lots of bookings for the Chemchem campsites. We avoided the place. To emphasize our distance, we'd put up a thorn fence where the path along the stream entered our compound. We tried to deter the tourists from wandering in and chased them out politely when they did. Even so, I was always interested to know who'd come to camp, especially researchers.

"Calm down, Jumoda," I said. "Who's at the campsite?"

He frowned, "I don't know. They told me they are researchers and do not have to pay for the village campsite. I think they're pretending. They should pay."

"What tribe are they?" I asked.

Jumoda admitted, "I only saw the camp staff; they're all Tanzanians."

I sighed, imagining the scene—Jumoda striding into camp in his pompous way, telling the staff they had to pay up. They would not like his attitude.

I suggested to Jumoda, "You may be getting irritated over nothing. Go back, wait for the presumed researchers to return. Legitimate or not, maybe they'll be willing to pay the camp fees."

Without another word, Jumoda strode off frowning, hitting his herding stick on the ground.

I was curious about the research group at Chemchem Springs, but I didn't want to show my interest by going to visit them in person. Jumoda had made it clear that he resented us being anywhere around what he considered "his" campsites or "his" tourists. To keep on good terms, and simplify our complicated lives, we avoided encountering campers.

To find out more about the mystery campers, I decided to write a note to the "researchers." I'd invite them for a drink at our place. I smiled to hear the slapping of another type of shoe, flip-flops. Gudo appeared.

We greeted. With an amused tone he told me his news, "The visitors over at Chemchem campsite, they are doing research. They are hiring us Hadza. Imagine, they have come to look at our feet!"

Gudo shook his head, a way of telling me that nutty foreign academics were

CHAPTER 14: THE FOOTPRINT MAN

back, doing their wacky studies. He sighed and said, "I'm going over to the camp now. The leaders have just returned."

"That's lucky for me," I said. "Please take this note to them and bring me back a reply. I'm inviting them to come over."

Flip-flopping back into the compound within an hour, Gudo brought me a note:

"Thank you for the invitation. We will be most happy to come to meet you at Mikwajuni. We'll be at your place at 5:30 p.m."

The note was signed with a name that sounded Tanzanian, Charles Musiba. Neither David nor I had heard of him. From the "we" in his letter, we reckoned his colleagues or assistants would join him.

Three men arrived in the evening, settling onto the cushion-covered benches

Charles videotaping people walking

on our small front porch. The most notable of the group was Charles Musiba—tall, dark, and handsome. His voice had an almost English accent, soft yet crisp.

Charles and his two colleagues introduced themselves as archaeo-anthropologists. Their main interests centered on ancient humans and their artifacts. David offered them drinks. We set to asking let's-get-acquainted questions.

Professor Tuttle, Charles' academic advisor, told us, "Chicago is our base. Even after 30 years, I still do not feel at home there."

Benedikt, who told us to call him Ben, wasn't from Chicago, either. He kept quiet and mostly just nodded while the rest of us chatted. Charles said he was in Chicago to do his Ph.D. at the university.

"And I'm not from there either. I was born here in Tanzania."

"Where in Tanzania?" I asked.

"Guess," he said with a smile and a lift of his chin.

In the best tradition of Sherlock Holmes, my thoughts ran like this: he was tall and slim in frame, had good bones, and was good-looking. He glowed with health, so I reckoned he must have been well-fed while growing up. His educational background was impressive, his social skills excellent. Best of all, he was articulate, with graceful manners.

I concluded he came from an educated, well-off family. Since the Tanzanians I knew with such qualities mostly came from around Lake Victoria, I made a guess, "Are you from Mwanza or somewhere around Lake Victoria?" He laughed and said, "You are right! I was born on Ukerewe Island in Lake Victoria."

While we sat on the porch, drinking our way through pots of tea and bottles of beer, we learned about their research. You guessed it—feet. Their job was studying footprints made by bare feet. Charles, Professor Tuttle, and Ben were keen to find out how shoeless people walked then compare their prints to fossilized footprints.

"Of course, you know all about Mary Leakey and the famous footprint trail she and her team found at Laetoli," said Professor Tuttle. He waved his arm toward the distant rift escarpment where the bulk of Oldeani and Lemagrut mountains hid the famous Laetoli footprint site.

We just nodded, not wanting to interject that we had visited Laetoli more than once. The first time Mary found an ancient footprint, she made a cast of my foot to study and compare. I was rather proud that my foot was a shape several million years old. That was before the Laetoli trackway became famous. We kept our attention on our visitors, wanting to hear what Professor Tuttle had to say.

He went on. "Well, three hominins, not the ancestors of Homo sapiens, but close, walked over the ash fall from Sadiman volcano. They left marks of their feet

Jeannette's foot beside a hominin track at Laetoli

impressed into soft ash. Then more ash and rain fell, embedding the prints in a kind of cement. The buried footprint trail endured for 3.6 million years." He paused to make sure we were listening. We dutifully nodded.

"We've been working there, studying those Laetoli footprints and making measurements."

Charles added, "And now we want to compare those fossilized footprints with modern ones. The problem is that modern feet are changed by wearing shoes. We want to look at the feet of people who don't normally wear shoes."

Ben had a chance to add, "We are taking measurements of the footprints of barefoot people. We think the Hadza are the closest we can get to people who don't cramp their feet into shoes all the time."

I took a look at the feet of our cluster. David's shoes had rubber soles harnessed to his feet by Velcro strips. I was barefoot, having kicked off my local sandals made of strips from old rubber tires. Our guests wore a variety of sports shoes with hot-looking socks.

When I checked in again, the conversation had walked on. I gleaned that Professor Tuttle supervised footprint research. He had a long-time interest in the evolution of human locomotion. Ben had completed a master's thesis on "fluctuating asymmetry," which he explained to David while I went to make more tea. All I could gather from the tail end of the explanation was that a correlation exists between physical symmetry and reproductive success. Balanced features beget better babies.

"And you, Charles," I said as I poured tea, "what is it exactly you are doing?"

"Why not come over and see for yourselves?" he replied with a big grin. All three seemed pleased with our interest. We felt a bit apprehensive about going to their camp, not wanting to get involved in Jumoda's fee demand. Despite that, we agreed to go in the morning for a first-hand look.

Our feet strapped into sandals, David and I dodged the three-inch thorns that the fever trees had scattered on the trail like caltrops. We strode along the dappled Chemchem path through the golden light of morning, eager to watch the Footprint Project in action.

Tents lurked here and there under giant fig trees surrounding the grassy clearing. Many of our Hadza buddies stood around or sat on the big tree roots. I recognized some who came from local camps and others who had walked from much further away. The Hadza acted as though at a reunion, happily clicking away in their unique language.

I greeted them with one of the most accessible Hadza words to pronounce, "Shayamo." With amusement, we noted a large pile of flip-flops and beat-up shoes at the edge of the campsite. Many if not most of these barefoot people were baring their feet for pay, not by custom or preference.

Charles came over. "Welcome. You came just at the right time. We are starting our work." He consulted a list and told us, "These are our code words for individual

The team studying Hadzabe tracks on paper

Hadza." He handed the clipboard to Professor Tuttle, who took the list and began to call each person from his or her roost under the big fig tree. Charles started the measurements.

One by one, each Hadza man or woman came to stand on a scale. Tuttle noted weight, then height. The designated subject then went to sit in a chair. Ben, using special calipers, measured foot length from the longest toe to heel, then heel, and width of the sole. Interestingly, sometimes the first toe was the longest, sometimes the second.

The professor wrote notes as Charles and Ben attached a plastic "compass" to the bottom of the foot using two day-glow elastic armbands. They rotated the foot in and out and gave a reading of the angle. Lastly, Ben held up each foot, and Charles took a photograph of the bottom.

The waiting Hadza looked on with their particular kind of glee at the idiocy of outsiders. Local people passing by stopped to stare. Most just looked puzzled or shook their heads, then went on their way. Jumoda arrived and stood around. We hoped he'd accepted this team, whether they paid the village fees or not. His initial frown faded away, replaced by interest. Ah, I thought, his natural curiosity has overcome his irritation.

After the measurements and photographs came the part I liked best— the paper trail! The three researchers set up a walkway. They used those old-fashioned perforated pieces of wide carbon print paper. Charles or Ben folded back a transparent plastic protector and peeled off strips of plastic-coated paper. Charles videotaped Hadza striding across this trail, like fashion models on a catwalk. The barefoot subjects left very nice carbon imprints of their feet as well as their stride. Everything would be measured later.

All the measuring, negotiations, instructions, and repeats wore out the researchers. The Footprinters looked tired out by the early afternoon. The various onlookers, including Jumoda, had become bored and left. Charles rewarded the Hadza for their participation with maize meal, dry beans, and cooking fat. I was glad to see the team selected Mzee Matayo, one of our oldest and dearest Hadza friends, to conduct the ritual of distribution. The sure-footed Hadza drifted away to retrieve their sandals, flip-flops, boots, and ill-fitting sneakers or plodded away on leathery naked feet. They'd return for more foot-printing next morning.

We invited the team back to our place for refreshments.

Sitting with a cup of tea, Charles told us, "After we collect samples from 50 barefoot people, we'll be moving." Already we felt fond enough of this man to regret his departure.

"Where will you be going?" asked David.

"We're headed over to Yaeda Valley, where Mabulla is now working." We knew Mabulla, another Tanzanian archaeologist who researched ancient living sites and rock paintings. Tanzanian archaeologists or anthropologists were valuable rarities. We appreciated the few who had found their way to the many places of interest in the Eyasi area.

Charles went on to say, "It might be a day or two before we finish here, then we're off. But we'll come back through Mangola on our way out." On the spot, I invited the team to a meal with us in the village before they left.

A week later, the Footprint Team returned, camping at Chemchem again. Jumoda had accepted them as legitimate researchers and left them alone. At our little party in the village, we got a chance to learn a bit more about Charles Musiba.

He spoke many different languages, a skill I envied. In his flawless English, he told us, "Of course, I can speak my home language, as well as Swahili and English, and I'm fluent in German and Greek." I asked him where he learned German. "Well, mostly when I was an undergraduate studying anthropology at the J.W. Goethe University in Frankfurt. Afterward, I worked on my master's degree in human ecology from the Free University of Brussels in Belgium." He added modestly, "Besides the European languages, I also speak about 10 African languages."

Most Tanzanians could speak a lot of different languages, but Charles' abilities surpassed anyone I knew; I was very impressed. He was exceptional in other ways, too. He had an unusual upbringing. "As the son of a radiologist in Mwanza," he told us, "I got hooked on bones. I would watch my father when he was looking at X-rays. They fascinated me."

"But you didn't go into medicine?" David asked.

"No, I was led astray. In secondary school, in the early 1970s, I apprenticed as a photojournalist for Tanzanian Information Services. One of my assignments was to report on an exhibit opening at the National Museum in Dar es Salaam. Mary Leakey and Jane Goodall were there giving talks. Meeting those two ladies inspired me, so I chose to study archaeology and primates, including humans."

So, Charles set his feet on the path that eventually led him to Chemchem and into our lives. Of course, he got his Ph.D. from the University of Chicago and has gone on to higher levels of responsibility.

He's a champion for education, especially for what students can learn in the field. When he joined the University of Colorado Denver in 2005, he immediately started a field school, taking students to do archaeological work in Tanzania. The program also helps raise public awareness in Tanzania and abroad about the value of archaeological sites and their deterioration.

We cross paths with Charles from time to time. Our favorite place to meet is during his field courses at Laetoli or Olduvai Gorge We've also found him in the USA and Spain. Footloose Charles gets around. We even met up to write a booklet on human evolution for Tanzanians called *Akili's Ancestors*. The star of the book is a boy from Charles' home area who goes on safari with his archaeologist uncle. We modeled his uncle on Charles. The boy is named Akili, meaning smart. They travel together to fascinating places in Northern Tanzania, where Akili learns about evolution, primates, fossils, archeology, and more.

What of his Ph.D. study on the barefoot Hadza? The Footprint Project ended up with 54 subjects between 6 and 70 years of age. Charles and his colleagues published their article in 1997. They confirmed that the feet of the Hadza were similar in basic features seen in the 3.6-million-year-old Laetoli footprint trail. What marvelous continuity there is on our long path, from barefoot hominins to *Homo sapiens* in athletic shoes!

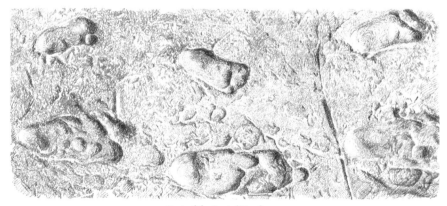

Part of the Laetoli trackway

CHAPTER 15
WILDLIFE LOVE STORIES
FOUR STORIES ABOUT OUR WILD NEIGHBORS: 1982-2003

Bird trapper with the box of parrots

We lived in a complex, beautifully woven, living tapestry along the Chemchem Stream. Bushes, boughs, and vines intertwined over and through trees. Tangled in the weave were bright birds, butterflies, bugs, and snakes. Hidden here and there in the shadows, we glimpsed bush babies, owls, and aardvarks, spotted antelopes and civets, maybe even a leopard. Bolder forms sometimes burst through—hippos, baboons, and monkeys.

This tapestry had a musical score too. Stand quietly and listen to the susurration of wind in the papyrus. You'll notice the burbling melody of the stream and the soft pit-pat of paws somewhere in the undergrowth. More strident notes are added by hornbills drumming, baboons barking, and a hadada ibis screeching. Above comes the song of doves purring and robin-chats singing so sweetly you could cry.

Put your nose into this tapestry and plunge into another sensory dimension. Sniff that civet musk; distinguish it from bushpig dung, and baboon poop. More enticing are the smells of spicy fig tree leaves, wild gardenias, and sweet acacia blossoms.

We loved this tapestry and watched unhappily as people steadily unpicked the strands. Through it all, we tried to cope with what loving wildlife entailed.

Parrots

Our eyes focused on the distant hazy horizon of the Eyasi rift as we drove down the Horrid Road one April afternoon. We were on our way home from an exhausting trip to Arusha. Our eagle-eyed assistant, Pascal, spotted two men walking along the road towards us. One carried a big box on his head. As we passed them, the man carrying the box swiveled it around in an attempt to hide what Pascal had already seen. "Ndege." *Birds*.

David stopped the car; we discussed what to do. We had no authority to question the men, let alone take the birds. But for a long time, we'd been frustrated watching the illegal trade. We could not just carry on as though we hadn't seen the tattered, feathered cargo jiggling and squawking inside the cage.

We stiffened our spines and confronted the men. They put down their burden, and we exchanged greetings. We peered into the box of plywood, sticks, and wire. Crammed inside, many bedraggled birds jostled and quarreled. We knew what kind they were by their bright green bellies and yellow shoulder patches—brown parrots. How did they get that boring name? Their more sophisticated name wasn't any better—Meyer's parrots, *Poicephalus meyeri*.

Brown or Meyer's parrot

We asked the men about their permits, where they'd got the birds, and where they intended to take them. Surprisingly, they admitted to having no licenses. They told us they'd trapped the birds along Lake Eyasi. Their destination was the notorious Dr. Shaka, at Prima Products, in Arusha. They even offered their IDs.

"How much does Shaka pay you for the parrots?" I asked.

"Sometimes 1,000 shillings each. We have 62 birds in here," said the tallest of the two, looking smug. Given that more than half the birds might die from heat, thirst, and hunger, these men would glean about 30 US dollars.

Since the bird trappers were open about their dealings, we decided that we'd try negotiating for the parrots' release.

"We want to return the birds to their homeland," said David.

They looked at one another with brief nods of alliance. You could almost hear their thoughts—two stupid foreigners right here, willing to buy them off. "Well, if you want to pay us Arusha prices, we're willing to sell."

CHAPTER 15: WILDLIFE LOVE STORIES   169

We quickly discussed this between ourselves. Yes, we decided to pay, something we had never done. We always worried that paying to free animals would only keep wildlife trafficking going. But putting our cash for these birds seemed the only way, short of physical violence, to start the pathetic parrots on their homeward track. In the end, we spent the equivalent of 20 US dollars in Tanzanian shillings. And that turned out to be a mere token of the amount in time and energy this rescue would cost us.

Part of the deal was that the trappers let us photograph them with the crate. We warned them with an empty threat that now they could be identified they would expect real trouble if caught again. Feeling a mix of relief, pride, and worry, we set off home with 62 new passengers screeching in the back of the Land Rover.

Sadly, bird trappers were a fact of life in the Lake Eyasi basin. Too often, we saw pelicans and storks strapped to the sides of an old pickup truck. In the back, stacked cages of birds, monkeys, or other wildlife sat in the blazing sun. We knew that many animals would die before being given water or shelter. The luckless creatures usually went to the exporters in Arusha or Dar es Salaam or smuggled boldly to Kenya via buses, private cars, and even taxicabs.

In the evening gloom, we reached home and set up an old tent inside our guest-

Parrot with damaged plumage

house. We collected branches for the parrots to perch on, then put in corncobs and figs, plus a big dish of water. As the captives cautiously emerged from their cramped cage, we could see their sorry physical state. Their wings were ragged from rubbing against wire, sticks, and each other for so long in a confined space. Stretching their wings was the first activity for most, then they eagerly fell upon the food and water, crunching up the dry maize kernels with their beaks like wire cutters.

The next morning, our little flock awaited us. We assessed the birds' condition first, then banded them. We had become experienced bird-banders while working with the East Africa Ringing Project. But banding brown parrots turned out to be an entirely new experience. Parrots' feet are sturdy, knotty, and very short; none of our rings would fit. That meant David had to put the oversized rings one by one in a vice, then trim them with a hacksaw. After that, we had to file off the sharp edges. Doing this 62 times cost us much of a day, along with four hacksaw blades and cut fingers, the definition of tedious.

We also learned that parrots could bite—hard! David put on leather work gloves and caught the parrots, one by one, grabbing them out of the tent. I held open a small cloth bag so he could thrust each snapping, struggling bird into it. I pulled the strings to close the opening and hung the bag on a rack until we could process it. Each parrot waited quietly in its pouch until its turn to be weighed, measured, and banded.

Banding was tricky. David held the first parrot firmly. Shrieking and flapping, it gripped glove or bag with both feet. Then it sank its powerful hooked beak through the leather into David's finger. He made a sound something like a lion roar, "FAAAAARK!!" He shook the parrot off.

Smothering a laugh, I caught the traumatized victim, put it in a bag, and brought it back. My chortling didn't help David get the parrot in a neck lock. Glowering at me, he gently but firmly clamped the split metal ring around the parrot's foot with special pliers. Getting my amusement under control, I recorded the parrot's ring number and measurements on a checksheet. The parrot squawked and bit David again.

Our curses, groans, and muttered numbers filled the air as we repeated this procedure 61 times, without the laughing. Then came the fun, releasing the bird. Most parrots crunched the hand that freed it. After each painful bite, we'd plaster more masking tape on the soft leather gloves until they began to look like some medieval knight's gauntlet. We released the processed parrots into their new temporary home—the bathroom of our guesthouse. We transformed the room into a large cage with branches in the shower stall, putting a food trough and paper on the floor.

The tempo of our task accelerated as the parrots in the tent began to chew their way out. They escaped all over the guestroom, making the job more difficult, so it took us a day and a half to finish.

Many pinches, bites, and expletives later, we had data: our population consisted of 20 adults, distinguished by their bright yellow crowns, and 42 immature birds. All the birds were in miserable condition. How much would it take to bring them back to full strength, to molt and replace their ruined flight feathers? When could we let them go? A bird expert told us, "Several months!"

We realized we'd committed ourselves to be caretakers. We had to keep the parrots in captivity until they recovered. And that course of action led to months of feeding them, releasing the ones that could fly. Eventually, we moved the crippled ones to yet a much bigger cage, one we built in our outdoor bathroom. There, we watched them die from black mamba bites, genet attacks, and sheer trauma when the birds fluttered and banged into the wire in fear. Each death made us depressed, frustrated, and angry.

Besides the costs we were incurring, we also caught unwelcome attention and criticism from the regional Game Officer. He accused us of interfering with his work, about trying to take wildlife matters in our own hands. He sent us nasty letters. We had just enough self-esteem to keep going.

The birds were actually worth some money if we'd been in the bird trade. I researched how much parrots cost: the trappers got about $1–2 per living bird. At local rates, that was a sizable amount for a little work. The trappers put sticky birdlime on limbs where birds roosted, prying them off, sometimes without their feet, and hauling them to a buyer in town. The town dealer sold the birds for about $7 to $17 each. The buyers of the birds, mostly in Europe or the USA, paid perhaps $500 for each Meyer's parrot. We had hundreds of dollars' worth of parrot stock, but we couldn't wait to see our charges fly free.

In the end, perhaps half of the birds survived. Slowly, they left us, flying into the trees and eating wild figs. We hoped they could now fly home. Luckily, parrots, love birds, and some other favorites of bird trappers were given temporary protection status by the Tanzanian government. Numbers began to recover. The whole parrot rescue episode put us off trying to do anything like that again. But our concern and love for wildlife caused us to get involved in other events.

Dikdiks

I have a special affection for dikdiks. My love affair started when we lived in Serengeti National Park, where these amazing, knee-high antelopes darted from one clump of bushes to another. Dikdiks are small, innocuous, and incredibly "cute," with huge, lashed eyes and a nimble, careful way of walking and leaping. They are adorable creatures with pointy noses, furry bodies, and sticklike legs.

They are one of the few monogamous mammals, living in pairs, protecting their territory against other dikdiks, and the predators who find them delicious. Raising baby dikdiks is a tough job, made even harder when humans interfere.

Dikdiks were also part of the Lake Eyasi landscape. When we first came to Mangola, an event brought home to me how vulnerable the dikdiks are. It started in darkness, as I lay on a blanket in a sunflower field, star gazing. The Great Giraffe was there, its head the Southern Cross, its neck bright Alpha and Beta Centauri, with the nose poking into a thick canopy of stars, browsing the Milky Way. Orion the Hunter stalked the sky, his dog at his heels. Lying on my back, I hunted for Halley's comet, getting closer and brighter each night.

I put my binoculars down on the blanket when I heard a car arrive. A door slammed, followed by quiet. But soon, the sound of Kaunda's rubber tire shoes on the hard ground signaled he'd come to interrupt my perusal of the heavens.

"Visitors," he stated. Kaunda, the farm's unofficial sub-manager, was a happy-go-lucky, competent camp man who had two other special skills—finding people in the dark and using very few words. I hoisted myself up, tossed the blanket over my head and shoulders, and set out across the field. Peering from behind some bushes, I saw two khaki-clothed figures next to an impressively new but dusty Land Cruiser. A man, and another I took to be his adult son, looked relatively friendly. I approached them.

"Hello," I said. "Can I help you?"

"Hmm, hello," said the man. "I am Richard Hunter. I have come to kill wild animals." Well, that isn't exactly what he said. But that is what he was—a hunter. He told me he worked for the World Bank and had the Christmas and New Year holidays off. His job didn't impress me, but his guns did, the open door of the car showing two rifles plus a pistol at his hip. He'd heard about Lake Eyasi. "People in Karatu told me there is still some wildlife here to shoot. It's outside protected zones, so we don't need permits."

I shook my head, not knowing anything yet about hunting permits.

Mr. World Bank said, "People told me that you know the Hadza hunters here. They'll know where the game is."

Game. I hated thinking of wild creatures as game. Indeed, hunting was a game for men like these. With their high-powered rifles, scopes, sights, cars, spotlights, and trackers, they held all the cards. The wildlife had no real defense. Mr. World Bank wanted me to round up some guides and trackers. "Me?" I asked, "Me?"

I wasn't really against hunting. My father had taught me to shoot, carry, clean, and use a rifle properly. I respected professional hunters. Regulated hunters in Africa who paid a lot of money for their safaris caused less havoc and needed far less in-

The dikdik slayers

frastructure than the kinder-hearted photographers and tourists. The fees big game hunters paid did provide some local income, and the extra protection to wildlife areas could be significant, too. But on the whole, I had no sympathy for people who just wanted to kill animals for fun, target practice, or trophies.

My desire to help these two had a temperature below zero. "Me?" I repeated. They looked at me as though I were drunk or deaf. They looked at Kaunda, obviously a Hadza. I looked at him, too, and asked in Swahili, "Do you think anyone would want to go with these men now, at night?"

I hoped the answer would be no, but he said, "Sure. I can get two guys from camp. They like to hunt and bring home meat."

After they left in their big fat Land Cruiser with their two Hadza guides, I paced around in the dark, wishing I'd sent them away. The pair returned the next day, grinning. I kept my mouth shut as they bragged about the zebra they'd shot. Then Mr. World Bank waved animatedly in the air, showing how he'd whipped out his pistol and shot a dikdik as it tried to flee. I turned away; the image made me cringe, and my eyes filled with tears.

A dikdik; he killed a dikdik. I held my lips together and clamped down my teeth, trying to stop the tears. To kill one of the dikdiks parents doomed the family—the mate would be unable to defend the territory alone, and the young would be exposed to pythons, leopards, hyenas, jackals, lions, and predatory birds. Dikdiks had enough enemies without Richard the Dikdik Slayer.

Afterward, I became adamant about not helping any hunters. But then a decade later, Matayo, our Hadza hunter friend, came with an orphan dikdik held in his arms. Matayo proudly admitted that he had shot the mother. "Nyama nzuri sana!"

Our first sight of the baby dikdik

Delicious meat. Yes, we knew dikdik meat was tender, and the dead mother probably had provided just enough to share morsels with the rest of Matayo's camp. For some reason, they didn't roast the baby but had brought it to me. Matayo held out the small creature— "Here, Mama Simba, this is for you."

With mixed apprehension and delight, I accepted the furry bundle. She was about the size of a small cat with legs like pencils and large dark eyes. A young mammal's sole aim seems to be to make its caretakers adore it. We quickly fell in love with our infant.

What was I to do with the nuzzling little thing? Feed her, of course. Milk first, then fruits and nutritious leaves and grasses. Diki-Diki nibbled them with delicate abandon. Days passed, and the bright-eyed creature went from sleeping in a box under my desk and bumping into things inside our house, to stumbling around our outdoor bath area, scaring the parrots. As she gained daily in strength and boldness, we let Diki-Diki get acquainted with the big outdoors. She freely romped about the outdoor kitchen area and our compound. We watched her fondly as she wobbled, stotted, and raced around on her spindly legs. Even the monkeys came to play with her.

Diki-Diki was almost fully grown when one night she did not come back to our house to sleep. We were proud of her independence and felt that she had a good chance of living in the wild. Sleeping out was a necessary stage in that process, but we worried, nevertheless. For several nights I slept restlessly.

One night, like any mom, I woke in fear when I heard my child's screams. A few plaintive squeals in the darkness, then an ominous silence. Rushing out with flashlights, David and I searched through the thickets—no Diki-Diki. We roused everyone around. People started searching, prowling the streamside, and crawling through tangled brush. No one slept much that night, and we didn't find Diki-Diki in the morning.

Dejection and disappointment followed as days passed and still no Diki-Diki. Finally, we had to conclude that the most likely scenario was that our resident python had swallowed her whole. We'd seen a python trying to eat another dikdik, but when I pounded on the snake with a broom, the dikdik escaped. We'd also seen a python squeeze to death and swallow a vervet monkey, so we knew it was skilled at catching small mammals.

Cycles of life, we reminded ourselves. We took no revenge on the python; it taught us respect as well as fear. In local folklore, pythons are guardians of the springs. One must expect them to claim sacrifices from time to time.

Hippos

My affection extended to creatures not nearly so attractive as dikdiks: I also felt bonded to the giant hippos. They had no fur to caress; did not dance or prance and were way too big to hug. The hippos are a longer and more complicated love story. I grieved when they died. Our neighbors and the game rangers were murdering them.

When we set up camp to start building at Mikwajuni, we woke up at night to hear the hippos munching. Finding them meant looking for black mounds in starlight or moonlight when their wet backs glowed in the dark. I'd watch the great beasts ghosting through the glade, their lawnmower mouths moving over the patches of grass, lips plucking, teeth cutting. They honored us with their presence. I loved these immense vegetarian monsters harrumphing and honking from deep in the papyrus thickets.

I found the hippos fascinating, but the villagers did not like them; they feared to pass along the trails around the springs at night. A hippo can be very dangerous if you are in its way as it hurries back to its home stream. A hippo's tusks can easily slice through flesh. But what stirred local hatred were big hippo feet tromping over onion and maize fields. The farmers' hatred literally poisoned the hippos. They began to disappear.

One day David shared the grim news of a big male killed. The hunters dared to come to our place and try to sell us the tusks. Another hippo was murdered soon after that male and in a most horrible way. Villagers put pieces of wood embedded with poisoned nails in the hippos' trails from and to the river. The hippos would walk on the nails and, in great pain, go back to the river to die in the water.

After David and I moved from our tent to a bedroom with walls, we seldom heard or saw the hippos grazing on our small stretch of open grass. Weeks passed without hippo signs or sounds. Then we heard some of their honks and bellows as newcomers conversed in the swamp, probably traveling in from Lake Eyasi's edges. My joy at the news didn't last long. On returning from a trip abroad, we got a report from Issa, the caretaker of our shamba, our large vegetable plot on the other side of the river.

"A hippo baby died in the water," he told me.

"What happened to its mother?" I asked.

Issa shrugged and raised his eyebrows in the universal I-don't-know gesture. When I asked Athumani about it, he said, "Yes, I heard the shots. The boys told me a hippo had been wounded and went into the water to die. The ranger couldn't get at it for the meat. The hippo drowned, and the carcass bloated up. It was floating just upriver and stinking horribly."

The news catalyzed us to write our unfriendly district game officer about the hippo killing in Mangola. Prompted either by the letter or by general reports, he came to Mangola to check on the hunting situation. He arrived at our place with a car full of armed men. In haughty Swahili, he told us that he'd come to sort out the killings and selling of wildlife meat. He went on to the village to warn and chastise the locals—no more killing of wildlife without his permission.

Presumably, permission meant allowing his local game rangers to "clear out the animals invading the fields." That involved shooting bushbucks, baboons, and bushpigs, as well as hippos. The local game ranger was a critical factor in the disappearance of local wildlife.

Months passed. We presumed all the hippos had gone. Then one day, David told me he saw hippo tracks by Chemchem. A mother and a baby! Our happiness lasted for only a few days. Sadly, there was nothing we could do to hide their presence. When a rifle shot rang out from the swamp one morning, we immediately wondered who'd fired and at what target. I sent the boys on reconnaissance. Jumoda brought the bad news. "Pole, mama," he told me sympathetically, "They have shot the hippo."

"Where?" I asked.

"At Chemchem, of course."

Angry and worried, I followed Jumoda to Chemchem. Before reaching the clearing, I heard and then saw village youngsters perched on the limbs of a big fig tree. They were chattering away like parrots, watching the scene below. A game ranger lounged against the trunk of a tree, holding his rifle. Several of our village friends were there too, with knives. It was a scene from a bad horror movie. I saw the hippo lying on the grass, twitching, and making bubbling noises, dying a gruesome death. More people came along paths from the village to watch and wait for the poor beast to die so they could chop it up.

Yes, it was the mother hippo. Her baby was already butchered, its pink meat piled on its skin under a tree. The mother was severely wounded. She tried to raise her massive head when someone slapped her on the side. I lost my restraint and shouted, "Leave her alone. Let her die." I felt so useless and impotent. Now I was exposing myself, too. Feeling as vulnerable as the skinned baby hippo, I slowly returned home. I sat on tree roots alongside the stream, waiting in dense thickets while people passed me, trying to get control of my feelings.

I made it home and hid for a long while in one of my secret spots. Meanwhile, the rest of our staff went over to Chemchem to look at the dead hippo. When they returned, I got up my courage to go to the kitchen area. They told me all about it, and I forced myself to listen. I shivered with horror, and David gave me a supportive hug. Even now, many years later, when I think of the mother and baby hippo, I brim over with despair, desperately wanting people to learn to share the earth with our fellow creatures.

Monkeys

We lost the conflict between hippos and humans but battled on. Arriving home after a long safari, Issa told us in rapid Swahili "Iko shida." *There's a problem.*

"Trappers came here. They captured monkeys by the river. We found one monkey with his arms tied behind his back. Somehow, he got out of the trap. But because his hands were tied, he couldn't eat. He was starving. His hands were all raw with huge sores. He stank. It was disgusting."

"What did you do?" David asked.

"We untied the monkey. We tried to feed it, but it was too weak to eat."

"Did it die?" I asked.

"Yes."

Issa stood with his hands clenched, deeply angry. He was usually an easy-going and friendly fellow. One of his chores was keeping local vervet monkeys from

Vervet standing bipedally

eating the fruits of our labors on our shamba. But here he was, defending them. The fate of the captured, tortured monkey upset him.

Issa told us more. "I didn't know these men. I went to the village office and told the secretary about these trappers and the dead monkey. He listened and sent the village militia over. And, Mama Simba, they caught them!"

"The men or the monkeys?" asked David.

"Both!" crowed Issa. "They caught three men, nine monkeys, and seven baboons. The militia destroyed all the traps; they told the men never to come back to trap at Chemchem or Gorofani village."

But the trappers did come back. We found traps along our side of the river and heard about them in other places. The demand for monkeys must have increased. Both vervets and baboons had value for medical research, circuses, pets, even as meat.

Jumoda came one day to tell us about other trappers close by our compound. The Mikwajuni vigilantes—Jumoda, Len, Sam, Gillie, and I—marched around the springs and over to the village to enlist the help of village officers. En route, we found the trapper with his pickup full of cages with monkeys and baboons. He produced a fake permit and sneered at us. That made us angry enough to hunt down our village chairman Julius at his home. The expression on his face when he saw us was, plainly, *Oh no—Mama Simba, and Jumoda, too. Double trouble for sure.*

He listened patiently, then said, "These men have come to the office and given the village a contribution of shillings, but I will check this out."

With great sufferance, Julius followed us back to the trappers and their sad cargo. We collected a couple of other village officials along the way. The posse stopped at the truck and looked at the monkeys clutching one another in their cages. I stood on the periphery as the village men took a look at the permits and the identity cards of the guys. They looked serious. Julius told the trappers to come to the village office in the morning.

In the morning, the trappers were gone. Later that day, I asked our chairman about the monkeys. The answer was less than satisfactory. "The matter is all sorted out," he told me. But of course, it wasn't. The matter remains embedded in human nature—greed, insensitivity, and ignorance.

Pori People
David and I and a few others wanted desperately to conserve the woven beauty of our wildlife tapestry, to stop the killing, cutting, and chopping. We couldn't deal with all the actors in the wildlife drama, but we struggled to do what we could. With the help of a little group of wildlife defenders, we developed a plan we called the Pori (wilderness) Project. The idea was to search the Lake Eyasi basin to learn more about trapping and killing wildlife.

The area we wanted to cover was roughly the area where the remaining Hadza people hunted and foraged. Nicholas Blurton-Jones, who studied the Hadza for many years, estimated that the area measured some 55 by 25 miles, a little larger than the Los Angeles basin, or about the size of an ordinary English county.

Unlike a sprawling city or a middling county, the Eyasi area is rugged terrain with many rocky hills, thorny defensive brush, little water, and countless places for trappers and poachers to hide. I hired field agents, namely Hadza hunters familiar

Hadza men hunting by Lake Eyasi

with the area to spy out the land. Local people knew the Hadza roamed all corners of Eyasi foraging for animals, berries, and honey. Hadza out roving did not arouse suspicions. The Pori People wandered the Eyasi wilderness and came back with verbal reports.

The Hadza spies produced some useful information about illegal activity and the abundance or absence of wildlife. The good news was they saw lots of wildlife. Even so, the numbers were far less than in the past. Plains animals like zebras, buffaloes, and wildebeest had greatly diminished. The Hadza spies encountered poachers everywhere. Some of the poachers killed animals for meat—a small proportion killed for trophies to sell. The majority trapped live monkeys and birds for profit.

Hunters, poachers and trappers were not the main killers of wildlife though. Farmers and livestock keepers erased animals that interfered with their crops or lives. Lions, leopards, bush pigs, and hippos could not exist in proximity to people. Most village officials, rangers, and local people seemed either apathetic or actively involved in killing wildlife.

We tried to take a philosophical view of the situation. Increasing numbers of people would inevitably come to Mangola, bringing more problems. Habitat changes would squeeze out not only the wildlife but the foragers and pastoralists. Conflicts over land and water were inevitable.

Protecting land is obviously the single most crucial element in trying to save wildlife and hunter-gatherers like the Hadza. Addressing the plight of the Hadza can also help save many wild creatures. The Hadza are few and live in an ever-shrinking range.

Awareness and appreciation of the Hadza now permeate the media. They have been featured in magazines like *National Geographic* and many documentary films. Luckily enough people now care passionately about these remnant foragers that they have banded together to work on land rights. Several organizations, especially the Dorobo Fund and The Nature Conservancy, have made an impact. Will land use for foraging, livestock keeping, farming, roads, and settlements be sorted out? If so, the wildlife tapestry in the Lake Eyasi Basin might remain intact, its multicolored, fragrant, and singing strands holding space for some of life's most fascinating creatures.

CHAPTER 16
A FOOL SAFARI
SEARCHING FOR DANIELA, A BUSH TRIP

David's morning coffee was essential. It was April 1st, the day for fooling around. So, I took out the fresh ground coffee we kept in a tin in our galley kitchen. I put in a note, "Sorry, all gone." I knew he'd be disappointed and look around for beans. I took them out, too, and left another note, "All gone." I hid the coffee beans in an empty jar and got out the grinder that often had some ground coffee left in its little drawer. I put another note there, "Sorry, all gone."

I predicted that by then, he'd be desperate enough to try the much less acceptable instant coffee. Into that tin, I put some freshly ground coffee and a note saying, "For real. Happy Fool's Day."

David was my target for the feeble joke because only he among us knew about the April 1st tradition of trying to fool someone. I quickly forgot my little trick because I had to get ready. I was excited about setting out on a local safari. I'd told David the week before that I needed to fool around in the bush. April Fools' Day was coming up, a perfect day for departure.

I loved trips into the hinterlands, and this safari sang to me. David expressed a bit too much happiness about staying home without me. But I just laughed, he'd keep things under control, leaving me free of home concerns. Two of my favorite Hadza friends, Gudo and Pandisha, were to be my guides and companions. We'd try new routes into the wilds, we had no time limit, and we had a focus, too. The ultimate aim was to visit Daniela Sieff at her research site, a vague place called Udachotek (oo-da-cho-tek). This mysterious destination was somewhere on the shore of Lake Eyasi, well south of Mangola. To get there would mean driving along the treacherous lakeside, across vast rutted cattle tracks, through the thorny brush, ditches, and soda pans. Ah, challenges to restore the spirits.

I never under-prepared for a bush trip, taking care to gather tools, tents, water, food, and an emergency kit. One part of this exploratory safari was to keep track of where we went, so I rolled up my precious maps of the vast area, putting them into cardboard tubes for protection.

Gudo and Pandisha arrived midmorning with nothing but themselves and the clothes they wore. I gave them tea and bread with peanut butter and honey, then continued gathering supplies. After eating, they filled water jugs while I put together food for the trip.

I saw David checking the car for me, feeling gratitude for his willingness and ability to keep our vehicles running. He made sure that all the automotive essentials were on board, like the hi-lift jack, tire irons, patches, and spare parts. I smiled when I noticed a steaming cup of what smelled like real coffee on the hood of the car. He didn't comment but glanced at me with a smirk. Uh oh, I wondered what sort of amusing revenge he might contrive for me, his Favorite Fool.

We three explorers set off to the south, driving through sleepy, dusty, onion-smelling Gorofani village. On the far side, we turned off to go to Mangola Plantation to ask the manager if he had any ripe melons. Since Johannes and Lena took over the farm, the variety of crops had increased. Their honeydew melons were my favorites. I grew some melons, too, and had one ripe, fat watermelon wrapped in a spare bedding roll. Alas, Johannes had no mature melons. Instead, he gave me some rather dried-out cobs of maize for roasting.

From Mangola Plantation, I backtracked to the crumbling bridge over the dry Barai riverbed, crossed carefully, and turned on a major dirt road leading west. I mentally closed my eyes to the desolation surrounding the ever-expanding villages. At last, we left the ugly shacks and dumpy buildings behind. A maze of livestock tracks replaced the road, threading through thick stands of commiphora thorn bushes. Thousands of hooves had worn hollows in the powdery soil, leaving ridges and hummocks. I swung the steering wheel back and forth, trying to keep out of the ruts and avoid the slashing branches thrusting through the open windows. We didn't want to close them because the humidity and heat was so hot inside the car. So we rolled the windows up and down, keeping dust and thorns out, letting air in, laughing at our exertions.

We slip-slid in and out of the eroded livestock paths all along the margins of the lakebed. The cattle that made such trails soon engulfed us in clouds of caustic dust. The Wasukuma, Irangi, and other Bantu tribes—or collectively the *Waswahili*, as researcher Wazaki called them—drove their cattle, sheep, and goats during the rainy season. That was when grass and water were available along the route. If they didn't sell them at the Mangola mnada, the monthly cattle market, we saw the herds going up the slopes to Karatu. If not sold there, the herders drove the livestock to Arusha, a long and dangerous way.

The fierce-looking Wasukuma cattlemen usually carried rifles as well as spears to deter rustlers. They could fight off any Datoga or Maasai who might think market-going herds easy prey. We tried to avoid the men and their banks-on-hooves plodding along. I steered the Land Rover out of the cattle tracks and bounced onto the gravelly slopes of the falsely dry-looking lakeshore. I had to take care not to get too close to the oozing crust with unknown depths of gooey alkaline mud beneath.

Slowly we headed west and south until we came to an expanse of glistening white soda. Usually, the pan was dry and flat, a great place to pick up speed and rush across to where the main track began again on the opposite shore. But because of recent rains, there were now slicks of water here and there, making driving a challenge.

If I slowed or swerved while crossing such a slick, the car could sink into the goo and get mired for hours, even days. I spied the almost imperceptible high ground between the slicks and the rock outcrops. Calming myself before the plunge, I fiddled with the cushions I had to use to let me see over the Land Rover's dashboard. I was ready. As usual, we talked in Swahili. "Brace yourselves," I told Gudo and Pandisha, "Hang on!"

The Land Rover fairly leaped off the bank and onto the soda shore. We zoomed along with a plume of dust scouring our eyeballs, coming in the windows and through the metal floor. Feeling confident, I pressed my foot down and picked up speed to make a big curve around what we called Poison Springs.

The springs were small seeps of warm water among rocky outcrops. As we drew closer, I spied white and pink flowers among the rocks, so slowed down slightly to look at them. They were desert roses. Their Latin name, *Adenium obesum*, referred to their stout stems shaped like bonsai baobabs with white trunks. They had voluptuously curving limbs tipped with large flowers. Horticulturists have taken these plants from the wild and hybridized them to produce a lovely palette of flower colors—orange, magenta, pink, red-striped, and solid white. They are sold in nurseries and gardens all over the world. Here, they were at home in the wild grandeur of Eyasi.

I wasn't the only one excited to see the blooming *Adenium*. "Mama Simba, simama, stop!" shouted Gudo over the roar of the engine. Hadza men are always eager to get at these bulbous plants. The desert rose's sap yielded toxic cardiac glycosides. The Hadza boiled the stems, and the resulting goop was an excellent arrow poison. Both Gudo and Pandisha got ready to leap out of the car. But I sped up, shouting, "No! I can't stop now. We'll stop on our way back, over there."

I pointed towards an island-like rock outcrop along the shore of the soda lake. On another safari, I'd discovered rock roses hidden there. Gudo and Pandisha pouted and shook their heads, trying to look annoyed. But they knew I'd do my best to keep my promise and settled back in their seats.

Desert Rose (Adenium obesum) and Jeannette

Accelerating on solid ground, I hit one of the slicks and slipped across it with an adrenaline-pumping half spin, fishtailed back onto soda crust and finally to the other side of the pan. We lurched up onto the wiry grass, and I breathed a sigh of relief. I realized that if there was more rain, we might have to find another way back to Mangola by the much longer route through the Yaeda Hills and Mbulu plateau. There was no other way along the lake other than through the soda pan we'd just crossed.

Careening in and out of the ruts, we wobbled our way south. The thorny brush got thicker; the rocky outcrops more numerous. Suddenly Pandisha called out, "Mama Simba, simama!" This time I stopped.

"There is a cave here, one that people use."

"Does it have rock paintings on it?" I asked.

"I can't remember."

Worth a try, I thought. I got off the main track and forced the car into the bushes, hearing the sound of the thorns scraping more paint from the sides. Ouch. I could go no further. The diesel beast growled, rumbled with a sound like a horse blowing at the end of a hard ride, farted, and quit. In the quiet, I asked my inevitable question, "OK, tell me the truth, is this cave far?" They both shook their heads. I tried again, "Can we walk there and back before dark?"

"Ndiyo, mama," said Pandisha. *Oh yes.* I looked at the sky. The distant reddish sun descended in the west, wrapped in a dusty cloak. It was on its way to shine on the Congo, then admire itself reflected by the Atlantic Ocean waters. It would be leaving us in the next couple of hours. Yes, maybe time enough on this fool's day for a little exploration. We set out towards the cave, pushing through bushes.

CHAPTER 16: A FOOL SAFARI

The sudden smell of dung made all three of us stop in unison. We stumbled upon an empty cattle corral overgrown with weeds. From the corral, we hopped a dry wash and started up into the rocky hill, climbing through the underbrush, Pandisha leading the way.

I was ready to call a halt, admit defeat, and turn back when Pandisha turned and waved. "Over there," he called. He led us to a huge rock overhang, just the sort suitable for rock paintings. Alas, only smears of old red paint remained, nothing that we could decipher. There were signs of recent occupation though—a cold fireplace with three stones, scraps of cloth, a carved wooden spoon. In a low warning voice, Gudo suddenly said, "Watu!" *People.* I heard them too, livestock and herders approaching the corral below us.

We weren't keen on encountering long-distance drovers with guns who might be planning to use the old corral for the night. Crouching among the boulders, we sneaked down the other side of the hill, circling quietly around the rocks to find the well-hidden car. I imagined us in an old cowboy movie, astute Apache scouts avoiding cattle rustlers or the posse sent to round us up!

Soon we were out of the bushes and back on the track. The sun had set. Many books will tell you that night drops down like a curtain at the equator. It isn't true. Even three degrees below the equator, we had an hour or more of twilight, enough to find a campsite. We bounced along until we saw another rock outcrop not too far off the road, with some acacia trees growing nearby.

I liked having a grove of trees because I could park the car under them, tie my mosquito net to limbs above the Land Rover, and tuck it underneath my mattress on top of the roof rack. My rooftop set up was a secure, scorpion-free, snake-free, bugless bed that let me look up through the canopy at the stars. After I rigged my net, I joined Gudo and Pandisha at the campfire where they'd spread out the blankets I'd given them. In addition to the boiled eggs and potatoes I'd brought, we shucked those maize cobs from Mangola Plantation. They tasted like roasted owl pellets, but we chewed them up just the same. Gudo asked me to play a tune on my filimbi (flute/recorder), so I happily tootled at the half-full moon smiling on our adventure.

As dawn pearled the sky, we were up and gone before the sun, herders, or livestock appeared. We made good headway, keeping as far from the soft lakeshore as possible. Over the roar of the engine, I barely heard Pandisha ask, "Do you want to see the campsite of Shoomooling?" I translated the word to myself. Shoomooling was Schmeling. I remembered that Leoni and Hans Schmeling had told us about their old hunting camp hidden in the hem of the hills along Eyasi's shore. I agreed, "Let's try to find it."

Here was another chance to have a little adventure. Swinging the car up the bank along the edge of the sandy track, I headed into the thorny brush and across sandy watercourses. Finally, we abandoned the car among scrawny acacias.

We walked along to the edge of an empty riverbed so vast it was almost a valley. In the sandy river bottom, we encountered signs of recent elephants—piles of dung, big tracks, and broken, chewed-on bushes. That the ivory poachers hadn't killed them all yet filled me with delight. "Good to see elephants are still here," I commented.

Gudo just looked at me; I could read his thoughts—What? Elephants good? Mama Simba, you are a foolish woman, we are on foot, elephants are dangerous, they can kill you. Both Gudo and Pandisha seemed especially alert as they gripped their bows and arrows and looked around carefully, their faces and bodies tense.

As we neared the far hills, Pandisha cut across the riverbed and up a bank into the long grass. He beckoned us to follow. We rounded some bushes and came upon a small clearing dominated by a giant baobab tree.

Ah ha, this was the fabled campsite, well hidden from travelers on the lakeshore, but with no view, totally enclosed by bush. I was a bit disappointed and tried to imagine the camp in former times, remote and strategic.

Pandisha said, "We camped here many times with Shoomooling. He hunted buffaloes. Zebras." I vividly remembered the times when Hans had passed through Mangola Plantation with black buffalo carcasses bleeding in the back of his truck. On this safari we hadn't seen any buffaloes or zebras, just a few gnus running away from the herds of domestic livestock. Maybe the wildlife hid in the hills. I hoped so.

The abandoned campsite made me feel both sad and glad, sad for wildlife becoming increasingly rare and glad the bush was taking over again. Back at the car, I pulled out my roll of maps as we drank water and ate sandwiches in what shade we could find. I tried to locate the cave we'd visited the previous evening and the old campsite. The maps were my treasure trove of landscape information. They were scaled 1:50,000, very detailed. I could find our general position and make little dots on the map. David called such mapping Genetta's Positioning System. We wouldn't obtain a handheld Global Positioning System receiver until five years later.

We got into the oven-like Land Rover and backtracked to the lakeside. Now on high ground, I could speed along. Eventually, we saw cattle tracks leading away from the shore to a bit of swampy ground where two Datoga men were watching their herd of goats.

"Which way to Udachotek?" I asked in Swahili.

The men stared at us, consulted each other, and pointed in different directions. One shook his head and said in good Swahili, "Udachotek is about five kilometers along the lake. Then you must go through the bush away from the lake. Look for a small track after a place with many short trees and a dry marsh. The track goes through the marsh."

Such detailed directions were welcome and certainly not what I expected out here in the wilds. We tried to follow the advice, and after a bit of wandering, we found a cluster of mud houses.

"Udachotek?" I asked a bare-breasted woman in a leather skirt, with coils of brass around her neck. She nodded her shaved head, unimpressed by this alien invasion. I drove around looking for signs of Daniela; her Land Rover would be unmistakable out here. No vehicle, but we did come upon some wells barricaded by thorn branches.

I remembered the place, practically unrecognizable now. When we first saw it in the 1970s, small springs and pools abounded, a water source for wildlife and humans. Now the springs were gone, just muddy bottoms to the wells the Datoga had dug. Near the wells I noticed mud furrows doubling as watering troughs for livestock. Plots of withering maize surrounded the area, enclosed by fences made with chopped down wild berry bushes. I grieved for those hardy food plants killed to build a barricade to keep livestock from eating the maize. The sight depressed me. I suddenly wanted to get away and head for the wild hills. But before we could do that, I had to find Daniela. She was expecting us.

Daniela was a special friend. As a British graduate student, she had chosen to gather data on the Datoga living along the Lake Eyasi shore for her Ph.D. project. Daniela had severe doubts as to her desire and ability to do the fieldwork necessary. But after some culture shock, tears, and personal dramas, she seriously committed to getting the data needed for her thesis.

Her work required spending many hours recording information about how Datoga households worked and who got what to eat. She needed to find out the income from cattle, the cost of medicine for cows and people, and the overall health of the society.

Datoga assistant Shangai with Daniela

Asking around, we finally found Shangai, the primary informant and local sponsor of Daniela's work. He directed me to her *dosht* or hut. We talked as we walked towards the low structure with walls of many stout posts planted in the ground and a flat roof plastered with cow dung and mud. He told me Daniela hadn't arrived as planned. He expected her any day. Any day out here in the wilds meant any time in a week or a month.

The hot, humid afternoon weighed me down like a suffocating blanket. Saying goodbye to Shangai, I returned to the Land Rover where Gudo and Pandisha sat in the shade of the car. We decided to leave and return the next day. I reasoned Daniela might have had car trouble, and if far away, she'd be late getting back. I left a note at her dosht, sticking the paper prominently between the posts at the entrance.

With a feeling of playing hooky, we three set off to find our way into the hills. We searched out the stony tracks heading up through thorn trees and long grass. Relieved and revived with cooler air and renewed freedom, we sang as we got well away from the Datoga settlement. We even laughed as we started to slap at the tsetse flies that signaled wild country.

I kept driving higher, hoping to intersect an old hunting or mining track along the crest of the hills. From there, we could look out on the terrain to the east, the wild area of the Yaeda Hills, a mass of broken and rocky granite. Tsetse flies and a glimpse of a bull buffalo signaled wilderness, and I breathed deeply with pleasure. Moonlight made dusk linger, helping us find a camping spot as the sun slipped across the mirrored slick surface of Lake Eyasi and over the rift wall. I made a pot of soup while we grilled our last cobs of maize, now about as palatable as pinecones.

Looking at the pink rocks highlighted by the sun's farewell afterglow, an idea brightened my brain: before we returned to look for Daniela, we'd try to find the hidden gong rocks again. Dawn's peach-colored clouds smiled at us as we got up. I stuffed some energy cookies in a bag and grabbed a water bottle. My companions merely strode off into the rocky outcrops holding their bows and arrows.

Pandisha led us steadily up into the hills through bush and along canyon bottoms. We were alert, looking for signs of buffaloes and elephants, walking carefully and quietly. I was wearing my lace-up tennis shoes and thick socks instead of my usual rubber tire shoes. Walking in the wilds in the rubber tire shoes had often left me with blisters and cuts. But the tennis shoes turned out to be a big mistake at this time of year. The grasses had grown with the rain, their little flowers becoming stickers. The twisted seeds from tangle head grass screwed their way through my socks, even the shoes, and embedded themselves in my feet. Pandisha's trousers also attracted skewering seeds, so he and I soon started picking away at the needles. Gudo flapped along in his shorts and flip-flops, chuckling to himself when he looked at us.

Midmorning, we rounded the flank of the highest range of hills and dropped down across a steep rocky canyon. Resounding barks of big, bad baboons echoed around us, and we hurried away, following the streambed, all our senses focused on sounds around us. We emerged onto the east-facing trough in the hills where several massive outcrops loomed. My heart filled up with a love that flowed over this vast, free landscape. Such wildness, I thought, as I scanned the rugged scene in front with its backdrop of distant purple mountains and bright blue sky.

We climbed to the top, delighted to look across to the open face of the gong rock. Like a klipspringer antelope, I bounced down the outcrop, Gudo and Pandisha following. Soon we stood on the smooth granite dome with its slices of rock pitted with holes, the gongs. Gudo and Pandisha both greeted the gong rocks with respect. I noted that Gudo held his bow and arrows in his hand and caressed the stone while chanting some words, like a prayer. Pandisha also ran his hands and arrows over the rocks. Were they seeking a blessing on their hunting or listening to ancestors who might have been there? Indeed, the setting had magic and majesty, as well as mystery. No one knew who and when humans pounded the dimples into the rocks.

We gonged on the rocks and sang songs springing from the spontaneous rhythms we produced. Then we moved to a high point on the outcrop to sit under some bushes while we ate cookies and picked the stickers out of our shoes and clothes. Gudo joined in, and we all enjoyed an ancient and soothing primate ritual—mutual grooming. Gudo looked around, then headed for a bush growing between some boulders. He and Pandisha selected a few long straight stems.

While they cut and trimmed their arrow shafts, I crossed the narrow valley to a rocky hill, intending to keep my promise and say hello to the cycads. The palm-like plants with their waxy dark green fronds looked like fossilized ferns. I stared at the descendants of plants that originated millions of years ago, tromped on and nibbled during the age of the dinosaurs. The cycad lineage was three hundred million years old. During that enormous amount of time, cycads evolved into many different kinds. I was looking at a species that developed in the last 12 million years, still a very long time! During that time, our human line started diverging from the chimpanzee line.

I tried to take in the wonder of being a mobile human looking at cycads stuck in their rocky fortresses. Always a lover of survivors, I wished them well in their struggle to leave descendants in this wild landscape, along with their increasingly rare companions, the Hadza people.

Gudo whistled, and I returned to the gong rock. Happily, we sauntered along the dry riverbed of the shady canyon. Suddenly a chorus of baboon barks and shrieks stopped us in our tracks. The sounds ricocheted back and forth in the rocks. Gudo and Pandisha dropped into a crouch, bows out, arrows primed. No, I whispered to myself, please don't shoot the baboons.

I only realized that the baboons weren't barking at us when I crouched down behind the men. Through the gap between their bows, I saw something that made my heart pound. A large leopard stood on the bank opposite, almost invisible in the shadows, staring at us. Its body was poised, a young baboon dangling from its jaws. With a ripple of spotted fur, it turned and bounded away, an apparition, gone in an instant.

I breathed deeply, trying to slow my heart, my mouth dry. The hunters seemed calm. They stood and let their bows swing to their sides. They chatted about the event in Hadza language as we made our way back to the Land Rover.

In the shade of the bushes, I took off my prickly shoes and let my feet loose, got out the map rolls, and laid them on the ground. I tried to locate the gong rock and put a question mark instead of a pinpoint among the twisted folds of the rocky hills. It might be wise to keep the site hidden, I thought, not only for the sake of the gong rocks but also for the cycads that were probably a target for plant collectors.

I drove out of the bush slowly, thorns screeching, adding designs to the already much-scratched car. Once out of the thorniest patches, I stopped under a shady tree. Now was the time for a break and my surprise—the watermelon! I unwrapped it carefully, feeling strange indentations on the smooth surface. I held it up, took a good look, and started laughing.

Gudo and Pandisha looked at me and then at one another, obviously thinking Mama Simba had blown some mental fuse. I turned the melon around so they could see what amused me so—David had carved a grumpy face into the striped green skin of the melon and the words, APRIL FOOL.

While they didn't understand the meaning of the message, they could see the cartoon face. Both my companions joined in with smiles as I laughed. I stabbed the melon with my hunting knife, slicing it into big chunks. We three sat on rocks gorging on the thick, juicy flesh, spitting the seeds as far as we could, thoroughly enjoying our silly game. Watermelon, the perfect safari refreshment.

Back at Udachotek, we found a note from Daniela. "Sorry to miss you. I was stuck halfway to Karatu. A lorry kicked a stone, and my windscreen dissolved into a thousand shards. I had to go back to Arusha to look for a new windscreen. I didn't get to Mangola 'til yesterday—David was home. I got out here this morning and am measuring the nearby households, about a five-minute walk. Get Dangaida to show you the way. I'll be back at 4:50. I've made arrangements! You have a call with David at 5 p.m., courtesy of my radio. It's finally working. See you soon."

Dangaida had striking circles of scars around her eyes. I gestured to her to lead the way and followed her swishing fringed leather skirt as we went along paths towards the wells and maize plots. At last, we came upon Daniela. Jumoda was there as well, in his research assistant role.

Daniela stood in the shade holding up a baby in a basket scale while Jumoda marked its weight on a checksheet. Taking my arrival as their signal to quit work, they handed the baby back to the mother. Daniela and Jumoda started collecting notebooks, baskets, and other gear and we all headed back to the dosht.

By now, it was close to radio call time. I could hardly hear David's voice, but I could tell him that we'd be out in the wilds another day at least. Radio duty done, Daniela said she had no desire to eat but was utterly exhausted. She headed to her

CHAPTER 16: A FOOL SAFARI

Daniela and Jumoda doing fieldwork at a Datoga home

tent on top of her Land Rover. I knew that she found her fieldwork hard—she had to use two foreign languages—Swahili and Datoga—as well as cover long distances to and from her research sites. She also had to handle Jumoda, a necessary, if sometimes irritating and unreliable assistant.

Free of obligations, Gudo, Pandisha, and I drove into the nearby hills where we could camp by ourselves.

"Ha!" said Gudo, "I'm happy to get away from those cows. All those flies, all that mooing and noise!"

Pandisha's verdict was equally strong. "For me, anywhere near a Datoga boma is bad. They smell bad, and there are more mosquitos, too."

Knowing that in the past, there had been lethal skirmishes between Datoga and Hadza partly explained their dislike. They also resented the way the Datoga had invaded the hills. The pastoralists fenced in the springs and let livestock foul waterholes that had once attracted the wildlife the Hadza hunted.

Halfway up the hill, I slowed to go around the fenced-in maize plots. Jumoda and Shangai were there. They greeted us and twisted off some fresh corncobs. Gudo and Pandisha stoically took the bundle of cobs while I said thanks and drove on to find a camp spot farther up the hill.

Moonlight beamed down on us while we roasted the maize. The kernels tasted sweet, juicy, utterly delicious. Gudo and Pandisha toasted their arrow shafts too, strengthening and straightening them. I took out my recorder and let the moon's song slide through me into the little flute. We slept well—no noises, mosquitos, fleas, ticks, or intruders, just the purr of nightjars. Or was it a leopard?

At dawn, we toasted more maize, drank big mugs of tea, then drove back to the lakeshore. Gudo and Pandisha said they wanted to go hunting upslope away from the Datoga settlement. They claimed they'd be back midday. Daniela, Jumoda, and I set out with lists of households and family members to visit. The routine of data collection entailed having to greet everyone and do the tiring, repetitive work of weighing, measuring, interviewing, recording everything in great detail. Well into the afternoon I managed to persuade Daniela to take a break. We collapsed in a shady patch. "I'd better find Gudo and Pandisha and head home," I said. "If you don't mind, I'll say goodbye now. See you at Mikwajuni on your way out."

The sun dropped like a distorted orange through a gauzy film of clouds towards the rift wall as Gudo, Pandisha, and I reached the edge of the big soda pan. As promised, I took them to where the rocks sheltered the desert rose bushes. They cut stems to boil later for arrow poison. I lopped off two tops because I wanted to plant the cuttings. A glance at the sky told me it was too late to drive home. One more night in the bush would be a last blessing on this fool safari.

Near the rock outcrop, we found some open ground for our camp. No large trees stood there so that I couldn't hang my mosquito net. Instead, I supported it with a hoe, a shovel, and my cardboard map tubes pushed into the edges of the wire roof rack. The structure collapsed in the windy night, but I had a reasonably good sleep despite the net hanging over my face. We got up before dawn, loaded our gear, and set off, not bothering to start a fire or eat.

Two hours later, we arrived home for a late breakfast. The men ate ravenously and left looking pleased, clutching the arrow shafts and bags of poison plants, blankets over their shoulders. I felt as though a good movie had just ended; I was hungry for a sequel, an encore. I thought that sharing another watermelon, and a trip report with David would delay the end of the tale. I went to get the maps to show him where we went. No maps! A sense of loss tore through me. The maps were in the tubes that fell off the car in the night. They were somewhere out on the lakeshore, probably trampled or used for fire starter, too far away to retrieve. Foolish me, but at least I could renew my memories by telling you this story.

CHAPTER 17
ROBBERS
THE EFFECTS OF EL NIÑO RAINS ON CRIME AND PUNISHMENT: 1998

Mzee Saidi at a flooded section of the Horrid Road

"Not even bones remained," Pascal told us when we returned from a particularly muddy safari. His words marked the end of a traumatic event that came with the unusual abundance of rain in Mangola. The storms of the wet season of 1998 were the work of El Niño, that switch in currents across the vast Pacific Ocean. Such events happen about every five years and this year, the global weather patterns brought exceptional rainfall to Mangola. Once again, we realized that we were not as isolated from world events as we thought.

The weather El Niño brought meant ruin to some in our pioneer farming community, fortune to others. The lucky farmers whose onion fields escaped the floods of December 1997 and January 1998 saw their crops rapidly appreciate. A bag of onions that generally sold for $15-20 to the truckers who came from Arusha was soon worth $100.

The increase in value for onions had two results. First, the rains brought a surge of prosperity to the onion-growing villages in Mangola. Villagers, including some very young men, bought trucks, fifteen in all. The pattern of commerce in the area changed. The previous system had outsiders coming in with mostly empty vehicles and leaving with cash crops. These new owners took their harvest to market themselves. They sold their onions or maize for a reasonable price and filled trucks with trade goods for the return trip to Mangola.

Secondly, the prosperity brought a crime wave. A trucker who used to arrive with $1,000 or less now had to bring more than $5,000. These were ideal conditions for robbers. Why not put rocks on the road to block a deserted stretch? And then, yes, hold the driver up and rob him! You could threaten him with a gun or even a rock or spear, or sheer numbers of robbers. Even easier was to mug the happy farmers who walked around with millions of shillings in their pockets. The Mangola villages had no police, no banks, no telephones, and few functional cars. Robbing locals was as easy as stealing fruit from trees on an abandoned farm.

Dubious characters flocked to Mangola from surrounding towns and distant cities. A lot of them took casual jobs such as weeding and picking onions while waiting for easier human pickings. A popular job for brawny young men was hauling onion sacks across the precarious Barai bridge, closed to traffic because of flood damage. The only way to get loads from one side of the river to the other was by tractor through the riverbed or by foot over the crumbling bridge. The immigrant porters monitored the payoffs with restless eyes. Other watchers decided to wait along the only major road into Mangola.

One of the first attempts at a hold-up turned out to be a bit of a comedy. A Unimog is a powerfully built German truck with a high road clearance. The Schmelings used one for transporting goods up and down the Horrid Road. Georgie was the Unimog driver, a neat, savvy young man who knew the road. One day he took meat from the Kisima farm to Karatu. He kept his eyes open, alert.

Georgie drove quickly, in a hurry to get back home to Mangola with his two companions. Dark shadows settled over the rocky road as they rumbled downhill, the Unimog chugging across washes and gullies on the descent from the highlands. They rolled down the long mountain flank that dives down a steep slope on the way to the lakeshore. Right there where you had to slow at the bottom, Georgie saw the line of stones across the road.

The bandits were waiting alongside their barricade in the dark, ready to pounce on their prey. They may have hoped their victim was an empty lorry coming to Mangola to collect onions. If alone, the driver was probably on hire and didn't mind if bandits stole all his employer's money, so they'd let him live and leave.

Alas for these bandits, Georgie was not an ordinary driver. He immediately understood why the stones were there and quickly steered the Unimog over and around the rocks. The heavy truck went sideways through a gulley, and Georgie drove away, leaving the bandits stymied. He didn't just escape; he went straight on to the Gorofani village office and reported the robbery attempt. Not content with that, he returned with a posse to catch the thieves. Alas, they had disappeared into the bush. The men pushed the rocks off the road. Georgie was a local hero.

CHAPTER 17: ROBBERS

Georgi's unimog meets a roadblock

Another event sank the goodwill in the community, leaving fear and hatred as debris. A farmer from a neighboring village collected money for onion sales, amounting to well over one million shillings, about $1,500 US at the time. He went to buy a soda from a kiosk in the heart of Gorofani village at midday. Two men approached from behind. One took out an iron bar concealed in his clothes and coshed the farmer on the head. They stripped the man of his jacket with pockets full of money and ran off, heading for the hills. Only women and children saw the event; they couldn't stop the robbers, but they could describe them.

Some villagers knew the thieves. The men had been working as porters for a while, carrying sacks of onions across the unstable river bridge. They were from Arusha, a town overflowing with the disaffected, unemployed, opportunistic, and cleverly devious. Villagers saw some of the young men coming to Mangola for temporary work as predators, prowling around in pursuit of a fast buck. They were known as *wahuni* (wa-hoo-nee).

The two wahuni who stole the farmer's money loped off into the wilds with the cash. They weren't the first to do so. One former British colonial administrator told us about a diamond thief he'd had to deal with 40 years earlier. "We convicted the bugger, but somehow he escaped and headed for Mangola. Thieves, rustlers, murderers, ran to wild Mangola country. No one could catch them. It was the Wild West of Tanzania!"

We reckoned Mangola was still part of the Wild West, but these two thieves did not go unnoticed. The keen eyes of villagers traced their escape route around the north end of Eyasi towards the Ngorongoro Highlands. Rumors filtered in that they'd stopped briefly at Endamagha, the remote village on the lakeshore below Oldeani Mountain. From there, the pair had allegedly walked up the rift wall on the twisting cattle track that led to Endulen village. People reported sighting the men buying goats from the Maasai herders. That was a smart way of turning stolen cash into cash on the hoof. Mangola had no police, let alone detectives, to track the trail further.

Not long after this robbery, a transport truck coming from Karatu was forced to a stop in the same place as the attempt on Georgie's Unimog. This time there were four bandits with a rifle and a pistol. Their first shot hit a rock. Their second passed through the back of the driver's seat. The people in the truck leaped out and rushed away into the bush.

The driver was glad his passengers had gone. He knew that he was the real target because he had the onion money. He jumped out of the cab holding his moneybag and flung it to the side as he ran off. The empty bag was a decoy and allowed the driver to run swiftly and safely away with the millions of shillings strapped to his chest—another hero.

The case of the two thieves who'd robbed the onion farmer remained a mystery, talked about often. People were angry that the two had got away and our crew and villagers kept us informed of developments. One evening, our village chairman Julius and another village elder named Kifua, came to ask our help. They'd had a tip-off that the thieves were at the cattle market in Endulen village in the highlands, selling animals. Julius and Kifua believed that the robbers would probably get a ride back to Arusha from Endulen with their loot, and if so, they'd have to pass through the Ngorongoro gate. The southern entrance gate was the only way vehicles could leave the Conservation Area on the main road.

We awaited the request for the use of our Land Rover, a frequent entreaty in such situations. We loaned it for jobs such as catching illegal charcoal burners or cattle thieves, or helping health officers do their rounds. But now we had plans for a safari. Surely, the village officials could get someone else's car. And how about all those new lorries or trucks that the onion entrepreneurs bought?

The two men looked at one another, stood, and looked at us, "Now, we want to get up to Ngorongoro gate today. We know that's where we'll intercept these two thieves. They won't get away from us."

Before they could ask, David told them, "We can't help with our car today, but we can help with money for fuel. You certainly can find another vehicle."

Since our own planned safari was to join a teachers' workshop in Serengeti National Park the next day, we made a further gesture of appeasement. To get to Serengeti, we would have to go through the Ngorongoro gate, so we added, "If you can't organize anything for today, we can take you to the gate tomorrow." They seemed satisfied with our offers and strode off alongside the Chemchem Stream back to the village.

CHAPTER 17: ROBBERS

The next day we loaded up and apprehensively went to the village office. We weren't entirely disappointed to find no officers around. Somewhat guiltily, we rushed away up the washed-out Horrid Road to the Ngorongoro gate where the guards said they'd not seen the thieves. After the time in Serengeti National Park, we went on a supply trip to Arusha town. It wasn't until two weeks later when we returned home that we heard the rest of the story.

The final act in the story of the robbers is a tale of local justice. You may find it disturbing, but it was undoubtedly effective. Here is what happened, according to the reports of various villagers and our home crew.

Cosh-man and Accomplice carried the jacket full of the onion farmer's money to the north end of Mangola. They walked up the rift on the cattle track. At a Maasai enclave near Endulen village they bought goats. Not being herders, they soon tried to sell off the goats at the cattle market. That took a number of attempts. But they didn't dare leave the Conservation Area via the main-road gate. To avoid such a critical spot, they hiked back down the rift to Endamagha village at the north end of the lake. Local people around Endamagha spotted them.

They knew the two men were the infamous *jambazi*—the Kiswahili word for robbers. How? One of the thieves was still wearing the farmer's jacket, and the other still had the bar that he'd used to smash the skull of the farmer. An angry mob grabbed the thieves and beat them up as they struggled to escape. The vigilantes tied them hand and foot and dispatched a man on a bicycle to Gorofani. He pedaled fast to deliver the vital news of the capture and demand a car to take the culprits to jail.

The news about the capture of the thieves resounded with proverbial drumbeats. From Gorofani, it rapidly spread to the home village of the farmer who still lay in the hospital in the distant town of Moshi. In our village, the executive officer and Jumoda, our youthful local vigilante, organized an event. They stopped one of the local *daladalas* (local taxi) and forced out the passengers. Jumoda and his backup made the driver take them over to Schmelings' place at Kisima and got a donation of fuel.

From there, the vigilantes rushed on to Endamagha in the commandeered daladala. When they arrived, there was quite a crush around the two jambazi. The two men were roped together, beaten so severely, they couldn't walk. Their captors bundled them into the back of the car. The leaders of the posse intended to take the captured men to Karatu police. But that was not to be.

During the time the thieves were being collected from the far corner of the lake, angry villagers assembled. Men, women, and children from all around the Mangola area swarmed towards the main road, by foot, bicycle, or aboard tractors. The mob wanted revenge. They did not intend to allow the thieves to "escape" to the safety of a jail. Everyone suspected that once in custody, the robbers could bribe their way out.

Our night watchman told us he could hear the mob voices from our house at

Mikwajuni. "It was late at night. I heard the car coming back from Endamagha, and I could hear a tractor coming through the village. It stopped somewhere near Mama Rama's place. When the car reached the village, I could hear the people shouting and screaming. I was worried. But I stayed to guard Mikwajuni," he told us with pride.

A villager, who was part of the mob, told us his version of the story, with a gleeful grin, as though the event was the funniest thing in years.

"A huge crowd, at least 500 people, stood there. There were young ones and old ones, men, youths, and women carrying babies. They blocked the road so the car couldn't pass. A big man stood right in the middle of the way. He lived in Jobaji where the injured farmer had his farm. This giant told the driver, 'If you take those thieves to Karatu police, nothing will happen. We won't let you pass. We'll wreck your car first and beat you up. We'll deal with these robbers right here!'

"We all swarmed around the car and forced the driver to drive into the middle of the village, near Mzee Kifua's house. Mzee tried to stop them, but there was nothing he could do. The mob pulled open the car doors and dragged the jambazi out. Everyone knew those guys from the Arusha market. We'd seen them here too, in their rasta hair and caps. The two men got tossed on the ground. Everyone started beating them, with clubs, sticks, feet, rocks, anything—even women and kids waited to take their turn.

"Then people started bringing diesel, petrol, kerosene. They poured the stuff on the bodies. They set fire to their long hair. Soon their clothes and everything was burning. One of them wasn't dead yet! He sat up and screamed. People beat him on the head, and more fuel poured on him. One person stood next to an onion lorry driver. He shouted in a loud voice. 'You, you come from Arusha. Be sure to tell people in Arusha that we don't like thieves here.' He turned to another driver and said, 'Selemani, you tell those people in Arusha what we do to thieves here in Mangola; they get burned up!'

"One of the jambazi had quite a lot of fat on him; he burned and burned. Then people started tearing down market stalls and bringing the wood, collecting branches, old tires, anything that would burn, piling it on the bodies. They kept the fire burning for two days.

"On the fifth, it was the big market day here. The lorries and people arrived, from Karatu, from Arusha. Everyone who came to the mnada went to look at the spot of the burning. Even the police came. They said, 'If you catch any more of these jambazi, don't bother bringing them to us. Just deal with them yourselves!' And they laughed. The District Commissioner came too, and he said much the same.

"When the fire burnt out, nothing was left of the robbers! Nothing, not even bones, just ashes! Usually, when they do this in Arusha, they put a tire around the guy's neck and burn him, but then you have a charred corpse. Here, nothing was left at all!"

CHAPTER 17: ROBBERS

We were shocked at the horrific story. Especially unnerving was the apparent delight at what our villagers had done to the thieves. Even demure and diffident Pascal, our man Friday, said, "After that, all the jambazi left Mangola. Even now, they don't stop here in our village; they don't get off the truck, they keep going. They know Gorofani village people are angry and will do something about thieves!" He nodded his head and said grimly, "Not even bones remained. Not a bone, just ashes."

Villagers watch the fire

Marabou stork

CHAPTER 18
A DAY ON LAKE EYASI
A SURPRISING BOAT TRIP WITH NANI: 1998

Flooded Lake Eyasi and drowned fishpond at Kisimangeda

The heavy rains of El Niño in 1997–8 filled the normally dry expanse of Lake Eyasi. At the edge of the lake, the Schmeling family had created several fishponds, fed by freshwater springs pouring out of the base of a striking volcanic outcrop. Selling fresh and smoked fish was part of their livelihood. As the alkaline lake water backed up into the fishponds at Kisima, Nani and Christian Schmeling tried to cope with the floods and save the ponds.

The couple had just married. Nani and Chris came home to Kisima, adjusting not only to the unprecedented watery world but to the recent death of Chris's father, Hans Jürgen Schmeling. Hans Schmeling had played a significant role in the Mangola scene—the owner of farms and estates, a planter and a hunter. His family were strong characters, too, and we loved them all—Leoni and her daughter Stephanie, son Chris, and the new member of the family, Chris's wife, Nani.

Nani was an adventurous Argentinian who had wandered across Tanzania alone, and into Chris's life. She was full of fun and charmed everyone, a free spirit who had acquired a unique "travel sense." Nani could tell incredible stories, and we enjoyed her company. She was not yet committed to anything except Chris and Kisima.

I hoped that she would become a friend, a companion to explore with, willing to share risks and adventures. As a first step, I wanted Nani to go with me in a boat on the newly filled Lake Eyasi.

One day, David and I got an invitation to dinner at the Schmeling's place, which opened an opportunity to ask her to go boating with me. As I walked the several kilometers over the hills and down the long sandy track to Kisima, I decided to wait for an opening and slip the idea into the inevitable conversation about the wonder of reborn Lake Eyasi. I hurried through the little village surrounding the Kisima farm, barely avoiding a dog that lunged at my leg. Rabid dogs were all too common.

I felt so good after my jogging walk and escaping the dog that I skipped through the open gate at the farm entrance. Skirting the herd of cows going home, avoiding people putting tools away and cars put under shelters, I heard the big generator roar to life. That meant the sun had settled down in the west, sinking behind the Eyasi Rift wall.

Right in front of me loomed the colossal rock outcrop that dominated this side of Lake Eyasi. I wanted to climb up to see the view from the top. I rushed up through the rocks and brush and stopped in awe. The new lake was huge, stretching from the flooded fishponds below to the distant purple escarpment in the golden glow of sunset. Off to the north, the blue bulk of the highland mountains stood at calm attention. I hurried down through the rocks to get to the house before dark.

The array of bungalows hiding among palm trees below was my destination. David, who had driven over in our new Land Rover, was already parked out front. I joined the brightly chattering people on the veranda, sitting at the massive table that Hans Schmeling's carpenters had made from palm trunks. With our backs to the stone-built house, we gazed down a grassy slope through palm trees to the lake, now a gleaming mirror of water reflecting the violet expanse of the Eyasi Rift wall. Drinking in the view along with a cold beer, we talked about the floods brought by El Niño rains, the drowned fishponds, the hippos seen in the lake again, the odd lack of fishing boats. Just the opening I needed.

"Nani, you intrepid soul, let's go for a boat ride! Just think, Lake Eyasi is a lake for the first time in decades. We can paddle along offshore and look for birds, check out the drowned trees, and see what's going on at the swamped farms. Are you up for a little adventure?" I paused, looking at her, hoping my enthusiasm was contagious.

"Well, I don't think we have anything planned," she murmured, seeming interested. She looked at Chris.

"What boat would you use?" asked practical Chris. I looked at David. He frowned a little because he knew what I'd say and didn't entirely approve.

I boldly said, "I'll bring our punt. It's not very maneuverable, but we can both paddle, and it should do fine as long as the lake's calm."

Chris glanced at David, who had put on a let's-humor-them look. Chris smiled wryly, then nodded.

CHAPTER 18: A DAY ON LAKE EYASI

"Tomorrow, then?" I asked.

"Sure, Mama Simba," Nani said and hugged me as we went in to dinner.

At dawn, David woke me before leaving for Arusha in our new car to collect his safari clients. After he left, I roused the boys to help me carry the heavy punt from the river edge and hoist it atop our old Land Rover. Len was the strongest and took one end, me the other, while Sam and Gillie each took one side. They looked pleased despite the early hour. With me gone, they could do what they liked—no boat, but a day to build a papyrus raft and go fishing.

I drove over to Kisima and found helpers to unload the boat at the shore of the lake. Next, I went up to the houses. The compound seemed oddly quiet without the early morning breezes that usually beat the palm leaves to and fro in a noisy rhythm. I tapped gently on the door of Nani and Chris's room. Early morning shadows flickered among the tall fever and palm trees around their house. I watched vervet monkeys stirring in the branches as I waited for a response.

Mumbles came from inside. Finally, a sleepy-looking Nani emerged. Chris came to stand behind her, nodded hello, and scanned the bright cloudless sky. I looked up too, into a beautifully clear and sunny day. Without a ripple of worry in my mind, I smiled broadly and told Nani I'd be down at the lakeshore when she was ready.

I went back to the boat to check the supplies: two paddles, and a bucket with a lid, just in case water got into the boat. In the bucket, I'd put a book, spare kangas, and a towel. I had a basket of food, jugs of water, a couple of beers, and a few cushions. I even remembered an umbrella to shade us from the fierce sun expected that day. Waiting for Nani, I sat on a stump near the springs and scanned around with my binoculars. Because of all the water, there were lots of birds, butterflies, and dragonflies, too. The unusually heavy rains of El Niño had drummed forth an amazing array of winged life.

Close to my spot, I saw warblers and weavers hiding among the papyrus fronds. Through my binoculars, I could see the birds further away. I spied waders and stilts poking their long bills into the mud at the lake edges. Further out, snowy white egrets strutted, their reflections following them in the still water of the lakeshore. The surface of the water was glassy smooth.

Some yellow-billed storks strode into view with their bright red faces and pale rosy plumage. I was delighted to see a spoonbill sampling the water with his odd-shaped beak. A congregation of blacksmith plovers in their black and white costumes clinked to one another as they explored the shore. Best of all were the flights of pink flamingoes. They ribboned across the hazy blue backdrop of the rift in the far distance like a loose sash of a sky princess. It was going to be a good day for birding.

Nani joined me, carrying a small satchel and a bag. "Ooh," she purred, "this is going to be fun." We pushed the boat partway into the water so we could climb in without getting stuck in the mud. Chris waved us off from the shore wearing a resigned expression that I interpreted as meaning, "There go two nutty women off

Birds on the lakeshore

to see the wizard of water." And there we were, floating merrily out into that vast soda lake. "Tutaonana baadaye," we cried out confidently. *We'll see you in a while.*

We paddled gently along the shore, heading north towards Oldeani Mountain, bubbly with delight at everything we saw. "Look at that view," said Nani.

I looked. All seemed pastel. I saw a giant painting, alive with color and light. The tranquil, shining water was silver with a dainty green hem of weeds at the shore. Grey green palm trees swayed in the breeze, ruffled by the wind. The greenery gradually melted into golden grass that blended into the purple of Oldeani Mountain. That old volcanic crater looked exceptionally beautiful, with mauve creases on its flanks and a topknot of greenish-yellow bamboo.

"Looks like a super huge dinosaur," said Nani. "Those ridges on Oldeani's slopes are like plates on its back."

The mountain dinosaur was more like a dragon to me, puffing smoke from the deep canyon on Oldeani's old volcanic face as early morning mist rose from the mouth. Human sounds brought my attention back to the shore. A couple of people waved and shouted greetings. We waved back. The pair sloshed along in the shallow water, heading towards a partly drowned hut that must have been far from the typical shoreline when built. Banana and papaya leaves also waved for our attention just above the lake surface.

The lakeshore looked so very different from what we knew. Now the water went into the palm thickets and flooded the cultivated shambas. These small farms provided the subsistence of many in the Mangola area. The farmers grew maize, beans, tomatoes, yams, and white potatoes but abandoned their crops as Lake Eyasi flooded over them. Heaps of rotting vegetation marked where they'd been.

The native palms and fever trees stuck up boldly from the water like sculptures. I thought it odd that they remained alive even though their roots were deep underwater. We paddled among the trees, looking for old nests on limbs and inadvertently scaring off roosting birds.

"Look there!" shouted Nani pointing to a thick crown of a tamarind tree. I turned to see a large bird hidden among the foliage, raising my binoculars just as it took to the air. The huge eagle owl flapped away into a patch of fever trees further away from us and the shore.

"Look Nani," I said, "its mate is still hiding in the shadows."

"Fantastic!" she replied, then added, "And over there, isn't that an osprey?" She pointed to a white and brown bird lifting off from a tree near where the owl had settled, causing an exodus of smaller birds from a nearby tree. Indeed, this was a perfect day for birds.

It was a good day for people, too. Individuals and small groups walked along the new shoreline, calling greetings: "Jambo, mnaenda wapi?" the inevitable *Hello, where are you going?* People always seemed to follow their hello greeting with a question. If you answered that question, you'd get another one, like "Where have you come from?" To avoid yelling questions and answers back and forth, we simply waved and watched people disappear into the palm groves or their disintegrating huts. The crudely built houses were mainly used during planting and protecting their crops. They were not a significant loss as they melted back into the lake mud.

"Why do you think there aren't boats or rafts out here on the lake today?" I asked.

"It does seem odd," said Nani "I don't have a clue why there aren't fishermen all over the place." We puzzled over this because fish had become numerous since the lake had filled up.

"Look at all the fish-eating birds," I said, pointing to the fleets of pelicans and the cormorants sunning themselves on the drowned trees.

"Their presence means there are lots and lots of fish in the lake now."

I hadn't thought to bring along fishing poles or bait; it was just as well. Nani and I felt lazy and calm, paddling along, taking our time. We worked out a slow and steady rhythm. I sat in the back, she in the front. Often, we just floated, going nowhere, admiring the grand landscape and scanning for birds. The water was so flat and the sun so bright we were dazzled. I draped a kanga over my hat for shade and gave Nani the umbrella.

We began to search for a place to rest and read, out of sight of people. Gradually we approached an inlet where feathery-leaved tamarind trees squatted in the water among the yellow barked acacias and palms. The trees welcomed us into a shady grove. No huts were nearby, just a long unbroken stretch of water between the huddle of half-drowned trees and the shore.

"What do you think?" I said, pointing at the pavilion of trees.

"Yes, it looks like the perfect place to stop."

We pushed in among the branches, tying the boat to a limb that hung just above the water. A gorgeous blue kingfisher with a striking red beak flew off with an offended squawk. The heat and humidity weighed on us, so we stripped down to our bare essentials. Feeling like well-hidden escapees from mundane life, we stretched out in the bottom of the boat with our pillows. We nibbled samosas and sandwiches, drank the beers, and read our books. Shade, quiet, and distant bird cries blended to make us sleepy. We napped a bit, rocked back and forth by soothing waves.

Waves? My mind rumbled into wakefulness. What had happened to the smooth, placid lake? We looked around and saw tree limbs swishing back and forth in the water. While we'd been dozing, a frisky wind had come skipping along the lake. The cool breeze was welcome at first, but as it rapidly increased in strength, we covered up with our kangas and shivered. We looked up through the waving branches and saw the sky clouding over. I looked at Nani; she looked at me.

"I think we better head back," I said.

"Right now!" she answered with some alarm and took up her paddle at once.

Going back meant paddling into the rising wind. David had designed the boat as a punt, with only six inches of freeboard. Our craft was built for use on a calm, sluggish stream; it had stability, but no speed. Our muscles strained to push us along. The wind continued to rise, making the lake water leap into the boat. We put our books and clothes into the bucket to keep them dry, but the lid kept popping off. We tried to grab the cover, but it sailed away like a frisbee tossed by a giant. We weren't worried yet, and actually enjoyed the challenge of playing with wind and waves. We paddled.

I decided to try for some help so crouched in the bottom of the boat and used the umbrella as a sail. The wind decided it had no intentions of helping so snapped the umbrella inside out and blew it away. Next, I tried using a pair of kangas as sails, cautiously standing up, bracing myself against the seats. That helped a little. But then the storm broke. Rain joined the wind. Waves started coming at us; rain and lake water whirled around us, stinging and cold. The bucket flew out of the boat.

Where had this storm come from so suddenly? We didn't have time to ponder the question; we just paddled like mad. It was slow and hard going. After a very tiring bout of paddling, we paused. I used my binoculars to peer into the rain. I could just make out the high rocks that marked Kisima. We headed for the promontory, skirting the groves of dead, thorny trees. The boat was filling with water, so we took turns bailing with our beer bottles, hands, and bags. The waves grew higher; more water splashed into the boat. Nani stood up, trying to move to a better position so she could bail. I could see this would unbalance us and shouted, "Nani, no, don't stand up!" Too late. The wind caught her, and over she went, grabbing at the side, pulling the boat sideways. It immediately filled with water.

Knowing my weight would only help the boat sink, I jumped overboard. We let the flooded boat right itself. Nani and I hung onto the side, dangling in the water. We watched our big bucket sail away, taking most of our clothes and books. The pillows and other gear also floated off or sank into the goo below. But the boat still floated, barely. We couldn't tip it back and forth to slosh out the water because it was too heavy. Also, we had no purchase on the bottom; the water was too deep to stand.

We turned towards land, towing the swamped craft and swimming as hard as we could. That meant having to dodge long-thorned fever tree limbs that slashed the lake surface. Equally dangerous were the drowned trees lurking underwater. We fought our way towards shore, avoiding obstacles as best we could.

The darkening sky shadowed the storm clouds; we stared into the wind and rain, trying to see where we wanted to go. At long last, we got close enough to the shore to stand in the mud. When I wiggled my toes in the muck, one of my sandals broke. Uh oh, I thought, now I'll have to tread with extra care. Luckily Nani still had her shoes on. We barely had enough energy to rock the boat, full of water and barely floating. Not enough water came out.

Jeannette and Nani return with the waterlogged boat

Having lost all our scoops, we tried to bail with just our hands. Standing in the water in our undies and clinging kanga cloths, we splashed out water as hard as we could. People started hailing us from dry land. We were aware of our near-nakedness but couldn't do much about it. At least the darkness hid us somewhat. The boat finally floated well enough so we could tow it further to shore along the lake edge. We were determined to get back to Kisima ourselves without begging help from onlookers.

And we did. Exhausted and laughing, we reached Kisima's shore at last. Darkness covered us as we pulled the boat onto the grassy bank and dumped out the water. Chris came towards us with a group of searchers, waving a light around to guide us in. He greeted us with relief, and didn't even scold us, much. But we weren't feeling guilty at all about arriving so late; we'd survived. So had the boat, as well as our sense of humor. We laughed at ourselves as we went up the slope to the house, our wet kangas flapping, happy to take warm showers before dinner.

Only then did I think about my precious binoculars. I put my hand on my breast and found the binoculars still hanging around my neck. They had taken a long soak in the alkaline lake; I was sure they were ruined. Peering through the eyepieces, I could only see a blur. Tomorrow, I told myself I would take them apart, clean, and dry them in the sun.

But they didn't recover, even after days of sun. It's hard to live in a lovely wild place without binoculars. I must tell you the comforting end to this story: another birding friend took those binoculars to the USA and mailed them to the company that made them. The company honored its lifetime warranty and restored them. I use them to this day. They remind me of that precious day of birds, unexpected wind and rain, the swamped boat, recovery, and waterproof friendships.

CHAPTER 19
THE MILLENNIUM MOVE
LEAVING MANGOLA: 1998 ONWARDS

Mud-map of our travels in 1998: Tanzania, UK, New York, Ecuador, California, New Zealand, Australia, Zimbabwe, Tanzania

The time had come, time to go. Stranded in six lanes of stalled traffic on a Los Angeles freeway, we finally decided where we would to go after leaving our wild home in Mangola. But it took five years to reach that decision.

In 1998, as the millennium neared, we reminded ourselves of our contract. We knew it was time to leave Mangola. More than a decade had passed since David and I agreed to build at Mikwajuni. And more than 14 years since we first came to the Lake Eyasi basin. We needed to face up to our decision to find another place, to force ourselves out.

Why leave wild fascinating Mangola, Mikwajuni, our handbuilt homestead and all the people we'd come to know? The main reason: we didn't want to become old in the bush. Bush living was acceptable as long as we were healthy and intact. But, no surprise, we were aging, becoming elders. We wanted fewer of the daily hassles that came with living in a remote Tanzania village. We wanted access to some basic facilities, decent roads, airports, internet, libraries, lectures, universities, plays, movies, and creative interactions. And most of all, we wanted to have more time with family before we became incapable of travel, and dead!

All those amenities and activities came at the price of tearing ourselves away from Mangola. Another cost was the fear of plunging ourselves back into the downsides of modern civilization: commercialism, capitalism, materialism, sexism, racism.

We loved our time in the Wild West of Mangola, but now we needed to go. Where? How could we best extricate ourselves? Starting anew in a different part of Tanzania didn't attract us. Other places in Africa didn't appeal either. Four world regions offered the chance to live nearer kin as well as being able to use the English language.

Our main choices seemed limited to England, the USA, Australia, or New Zealand. These all had family members but drawbacks too. Imaginative friends suggested other places we should consider, such as Costa Rica and Spain. Mobility and choice can be a curse. How to decide? To ponder our future, we retreated to our conference center, the rooftop of our bottle house.

Up on our flat studio roof, I set up mats and cushions, checking for scorpions under and around. Sharing wine or juices, we soaked up the ambiance, watching bushbabies peering from their tree-holes as the evening gold faded to deep blue. The sweet fragrance of acacia blossom mellowed us further while we watched stars poke through the fabric of night. On one night, a skein of flamingos honked across a crescent of moon, heading towards other lakes nestled throughout this grand part of the Great Rift valley. The dancing light and smell of the cooking fire in our outdoor kitchen highlighted the sound of human voices, as friends played cards at the communal table. What a wonderful world!

Indeed, it was wonderful—and so hard to focus on leaving it all. David and I would go round and round with ideas, trying to face up to how to depart. On one of our where-do-we-go meetings, I mused, "Why not go around it?"

"Around what?" David asked.

"The world! We can get a good deal on flights if we go around within six months. So long as we don't backtrack, we can stop over where we choose, add on side trips if we want to."

I watched David turning the idea over in his mind, as if looking for a bruise on an avocado. Finally, he responded.

"Well, that's not a bad idea. It would give us a chance to feel the vibes of different countries. Ah, decisions. We'll need to decide where to visit and when. Lots of people to write and maps to check out."

"I've already decided when to go," I said. "Do you remember what happened 25 years ago? Think!"

"Ummm...1973...Oh yes! We got married!"

"Yup! And I want our anniversary to be somewhere exceptional!"

And so, we planned our biggest safari ever. This was 1998, the year of El Niño, that year that changed so much in Mangola. The rains flooded the onion fields, filled the Eyasi lakebed, and altered our local economy. It was also the year that terrorists bombed the American embassies in Nairobi and Dar es Salaam. The world seemed full of situations that could change life entirely.

We decided to start our journey around the planet in September, our anniversary month. We wanted to visit new and old places, meet new and old faces. We'd

CHAPTER 19: THE MILLENNIUM MOVE

The yacht Cachalote at sunset, Galapagos Islands

go to all four of our possible resettlement sites. In addition, we chose the Galapagos Islands as the jewel in the crown of our around-the-world trip.

Hendrik Hoeck, our friend and colleague from research days in Serengeti, had lived for a while in the Galapagos as Director of the Charles Darwin Research Station. He knew the islands, as well as Spanish, having been born in Colombia. He agreed to arrange and guide a special tour. We still knew him by his nickname, "Pimbi," the Swahili name of his study animal, the hyrax.

Pimbi hired a sailing vessel, the Cachalote. Other friends and acquaintances joined us. We celebrated our anniversary onboard, a high point in our lives. We were aglow with happiness as ephemeral as the dewy dawn on a hot desert summer day.

Months later, at the end of our six-month trip, we flew from Australia back to Africa. We were on a roll and had decided to explore Zimbabwe too, an unplanned stop. We joined Pimbi and his daughter, Paquita, for a magnificent safari around that conflicted country. That visit was an excellent re-introduction to living in Africa. We'd explored more countries than we set out to and enjoyed almost every minute. From Zimbabwe, we returned to Tanzania.

Back in Mangola, we sat on the roof of the bottle house and conferred. We were still adamant about extracting ourselves from Mangola but now, in addition to *where to go*, we had greater impetus to decide *when to go*. Already, 1999 had crept in on us, and we wanted to leave in the next year, the millennium. It took longer.

During our round-the-world trip, we'd slipped into travel neutral, a free-floating mode, surfing events, living day-to-day. When we got home, we had to force ourselves to plan, to discuss, weigh, evaluate. Roof time discussions came more often. As the sun went down and the moon came up, we'd talk about options—so many choices, so many opportunities for sidetracks. We drifted.

Baboons on bottle house, Mikwajuni

The great Y2K fear at the passing of the millennium sputtered out, and the century ended. The world didn't. And we stayed on at Mikwajuni, still deliberating. From our wild office on the roof, we scanned around with sadness, knowing we had to leave this unique oasis; no one could get us out but ourselves. We listed all the pros and cons of each place we seriously considered going, then weighed the responses. Reluctantly, we had to admit that the US led. That decision smoldered while another year passed. We re-considered our choice, and the US still won out.

Choosing to live in the USA meant yet another step: where, precisely, in that vast nation? The millennium year went by while we researched places to live, narrowing them to the American Southwest: New Mexico, Arizona, or California. Those states had towns that would meet our needs, family members, and also intentional communities where we hoped we might find like-minded people.

We couldn't decide on a place without exploring in person. More planning sessions, then we booked our tickets. We left Mikwajuni in September 2001. David and I headed to Arusha to visit friends before catching our plane to the east coast

CHAPTER 19: THE MILLENNIUM MOVE

of the USA. Since we lived without television and news reports from radios, we were unprepared for the shocking news in town—two airplanes had crashed into the Twin Towers in New York.

We were stunned, along with the rest of the world. And guess where our plane was due to land? New York, just a few days after 9/11. We arrived at a strangely deserted JFK. Later we went to see the smoking site of destruction. National Guardsmen patrolled the area, and many sad faces looked at the disaster. People gazed at the photos of those missing and wept as we moved along, ushered by police and firefighters. The desolate scene touched us deeply, so far removed from our rural Mangola village. We wondered at the level of anger that caused men to kill themselves while destroying innocent people. There had been tribalism in Mangola, but we'd never before felt this degree of racism, hate, and religious fanaticism, alas, on both sides.

David and I flew from the East to the West Coast and took a careful look at places, spending several weeks roaming the southwest USA. We researched our way around, staying with friends, family, and virtual strangers, observing alternative living situations. We felt like refugees coming to a new country, apprehensive about where to settle. There were good and bad things about every place. We found it hard to commit.

Two days before leaving the US and still dithering, we ended up in a rather ludicrous situation. It helped push us to a decision. We were on the northwest edge of the vast metropolis of Los Angeles. We aimed to have a Thanksgiving family reunion at my sister's on the east side of the city. Stranded among thousands of cars, buses, and trucks on a crowded freeway in southern California, we stared ahead at the traffic jam. Smog and the funk of frustrated, apathetic, or furious people spread out like a gas attack.

Being stuck on that freeway was far more horrible than any trip on the Horrid Road between Mangola and the outside world. We decided that we just couldn't stay in the traffic anymore, got off the freeway, and made phone calls. Humble and full of gratitude, we went to stay overnight with some safari acquaintances. That forced us to make up our overburdened minds.

The final decision: we would buy a house in a new intentional community in Arizona called Milagro, meaning "miracle" in Spanish. It was our favorite out of all the places we'd seen. We called the developers of the community, and they accepted us. Finally, five years after starting our search and fifteen years after we'd agreed to leave Mangola, we had a place to go.

Following Thanksgiving, we returned to Mangola. We needed to wind up our business, loosen our bonds, and, most important, decide what to do with our hand-built homestead called Mikwajuni.

We resumed our bottle house rooftop meetings. On one memorable occasion, a scorpion stung me as I was laying out the mats. Stings were always a reminder that life was full of surprises; plans could and would go topsy-turvy. Joys balanced against traumas. During the agonizing rooftop discussions, we got depressed. We planned to leave; we even had a place to go. But what to do with Mikwajuni, the buildings, and the people?

The villagers had recently given us another 10-year lease on the land. Friends advised us to try to get an even more extended contract. We knew that would only bind us tighter to the place. Deciding how to leave gracefully was a real challenge. One night we finally agreed on our list of options:

Give Mikwajuni to the village.

Give over our compound to a local organization, like a school.

Set up a trust or non-profit, one that would let the Hadza or other tribal leaders use the buildings and compound as a visitor center.

Walk away and leave the place to human vultures and the ravages of time.

Tear it all down. Return our plot to the spirits of the oasis.

We re-read the list; I felt tears well up. I didn't want to face all the maneuvering that would be needed to organize a take-over by the village or any other organization. We were both weary of discussing, miserable about leaving our place, the people, and our wild world. It took the roar of onion lorries on the main road to bring us back from despair. No place could be perfect; everywhere had the equivalent of scorpions, trucks, traffic, and bad roads. We began to laugh at ourselves. All our weaving of plans would probably unravel as Africa, life, and the powers-that-be did as they wished.

At that time, we had a steady stream of visitors coming to see us. They came to enjoy our place, say farewell, and relieve us of possessions. This was one of the best and worst of times. Friends were a joy and eased our pain. Saying goodbye to places and people required emotional strength. Making arrangements for staff and dependents caused grief. We did the best we could, remembering a motto: Keep an open mind. Be kind.

Now the human scorpions came scuttling out from under their rocks. People came furtively to warn us that "certain persons" still wanted to take over our place. "They" wanted it for a lodge; "they" wanted it for their base of operations; "they" just wanted us to leave. We reckoned that the place we built and the land it was on ultimately belonged to the village. What did the village council want?

We had several meetings with our village chairman and elders. They dithered and avoided making decisions. We had difficulty seeing anything clear in the muddy political waters. We felt distant from the disputes. However, a growing wave of greed and power struggles soon forced us to pay more attention. Politicians in the district began to take an interest in what was going on. Politics were what pushed us into Mangola in the first place; now, politics came to help push us out.

Our situation began to get complicated. Politicians of the two Tanzanian political parties began maneuvering around us. The new political party wanted to enhance local village rights and was supportive of local initiatives. The old, entrenched political party wanted to increase centralized power.

District officials of the ruling party decided that the Chemchem water was under their control. An excellent way to demonstrate their power was to demand we move our houses further away from the stream. We refused to move cement block buildings at the command of the district officer. We went to the village council. We told them to stand up for their rights; the land was theirs to control; our houses belonged to the village.

In turn, the village elders told us we should stand up for our rights and keep our place intact. "You have a lease," they said. "Defend yourselves. Stay here." As village leaders shrank from the fight, outsiders became more interested in the conflict. Members of the new political party saw the underlying issue as the rights of villages versus districts. They took the side of village rights, a popular cause.

Lawyers wanting to support village rights offered free counsel. They donated time and energy to come to Mangola to bolster the cause. The whole mess became widely known as reporters began to write about the situation. A well-known reporter came from Dar es Salaam to talk to people. Our place became the focus of friction between the two political parties. Officials told us that a delegation would be sent to Mangola to find out who was pressuring us and whether or not to take action against district politicians.

During this political seesaw, we continued to struggle for a solution about what to do with Mikwajuni. We contacted schools, churches, hospitals, and non-governmental agencies, but none of them wanted to get involved in the political fracas. As the deadline came for us to leave, we simply had to decide. The conflicts and intrigues had gone far beyond the people who merely wanted our plot for tourism; we had become public property. We wanted to disappear. Before the governmental delegation arrived, we made our final decision about Mikwajuni—we would demolish our houses and go.

The demolition began with our home crew taking out windows, doors, shutters, and roofs. David and I moved our few personal things from Mikwajuni to a room offered by our dear neighbors, Chris and Nani Schmeling. They provided us with a base, meals, and constant loving support. We gave things away, sold stuff, took valuables to sell at a fair in Arusha. When almost everything was gone, and the houses were empty shells, our crew started taking down the blocks that we'd so patiently and lovingly made. We tore out the plumbing, filled in the carefully built septic tanks, and burned piles of thatch. The guesthouse went first, then David's studio, then our house, then the kitchen house and staff houses.

There were lighter moments. One happened when we took apart the metal rondavel that had been used for the visiting researchers. We unbolted the panels and stacked them up carefully. The buyer of this heap of metal piled the panels all-which-a-ways in his car, revved the motor, and beeped his horn to signal he was leaving. We watched in awe as he sped off upslope, the panels sliding out one by one, banging and clanking.

I raced up the hill calling "Stop, stop!" The man didn't seem to realize what was happening until he swerved onto the main track, and the last of the panels flung themselves to the ground with a crash. He got out of the car, stared in amazement, frowned, slapped his hat on the ground, then stomped on it. We laughed until our sides hurt.

Not so funny was to see the windows, doors, walls taken away, and the concrete floors pounded to rubble. We smashed up the rock fountain I'd built for David's birthday as well as the carefully made concrete tables and benches our builder Kefti, and I had crafted. Our outdoor courtyard-cum-bathroom came apart. We disposed of it as well as the outhouse up the slope. Finally, we pulled the staff house down. The only building left at the end was our studio.

In our last days of demolition, I'd remain at Mikwajuni and sleep on the roof of the bottle house. Inside were the smooth worktables David had made. I'd often stretch out on one of the remaining counters where we had done much of our writing and artwork, maps and displays. Sometimes after a long hard day of tearing our home apart, I'd walk back to the safety and sanity of Kisima with a heavy and angry heart, unable to eat or sleep.

David and I grimly kept on with the disposal of things, selling or giving away beams, lumber, tools, equipment, dishes, buckets. Trucks came and went. Aadje, our dear friend from Ndutu Lodge helped relieve us of many useful things. After the staff house went, so did all our employees except for Gwaruda, the watchman. That stalwart man with the long face and gentle manner slept propped against a tree or on the steps of the bottle house. He had narrowed his role to the protection of Mama Simba, the foolish woman who still slept among the ruins.

Mikwajuni disappeared, block by block. I found it especially hard to tear down our studio. Of all our buildings it was the most precious. It had taken the longest to

Jeannette in guest-house ruins at Mikwajuni, October 2002

build but doing so had been the most creative and fun. There it was, with its views from many windows, the big giraffe mural on one side, the wonderfully glowing designs made by the bottles. It hurt to destroy the roof, where we watched the reassuring sweep of the solar system, enjoyed the wild tapestry around us. From the top of that building, we had made soul-wrenching decisions about leaving. But it had to go.

The bottle house dissolved in fragments of cement, glass, and shards of sadness. The second floor and stairs came down; I could no longer sleep on the roof. We balked at wrecking a two-story tower with a concrete ceiling and a thousand bottles set in its concrete walls. We wanted to leave it as housing for our insect friends and a token of resistance for our human scorpions. But finally, the bottle house went. Each building's destruction was a painful sting.

The Tanzanian government sent the commission of inquiry. Men came to investigate what was happening. We received notice of their visit only at the last minute. Mikwajuni was a ruin. We'd booked our flights; we were soon to leave. The commission could do nothing to save the situation now. Various local, district, and central government officials came to where we stayed at Kisima with the Schmelings. The officials pulled me into the meeting, addressing me by my local name, Mama Simba.

They asked me to explain the situation. How could I possibly explain it? I tried to make them understand we were out of the picture. They were sympathetic, expressing their concern that the district was forcing us out. They told us we had a right to protest. I tried to tell them we'd already decided to go; it wasn't the pressure from district officers. The village officers should be the ones to protest.

My Swahili was not adequate for the political, social, emotional discussion; words wouldn't come. I fell silent, embarrassed to dissolve in tears. Finally, David appeared, and I disappeared to hide among the palm trees. David gave a short speech. He told them we were grateful they'd come, but they were too late to sort anything out. We were going. Maybe they could salvage some lessons for the next time such a situation arose.

And thus, we chose to suffer those stings of outrageous fortune and fly beyond the oasis.

Courtyard wall fragment - Mikwajuni - January 2020

EPILOGUE
WHAT NEXT? 2003 TO THE PRESENT

Of course, we didn't just fly away from Mangola and never go back. We have visited several times. After all, we had to check up on people we care about: Athumani and family, Gudo, Nani and Chris, Mama Rama, Julius and our nursing student Ruth. She always welcomes us at the impressive clinic she built on a hill to the south of the village. We've even seen Jumoda, who invited us to his family home for tea and showed us Tomikawa's monument stone. Sadly, Abeya, Matayo, Kampala, Saidi and Kaunda have all died.

It was Nature that took over our demolished homestead at Mikwajuni. Rising swamp levels drowned most of the big trees along the Chemchem Stream, but new trees flourish on higher ground among the ruins. The village government designated the Chemchem Springs a protected area. Vervet monkeys and baboons still harvest the fruit of the enduring tamarind and fig trees and we trust there are still bush babies and lots of bright birds.

Go visit, stay at the Kisima tented camp or Jumoda's campground or any of the many private campsites in Mangola. There are even new lodges built for the tourist swarms that we dreaded would eventually come. If you compare the vista from our former home to what is there now, you will be as astonished as we are at the transformation of the landscape. Like the onions grown there, many new layers of change have been added, huge onion farms where once was bush, the noise of pumps bringing water to fields, secondary schools, shops, houses, and hotels.

The muddled mix of tribes is even greater, the sprawl and squalor more noticeable, as is the diminished natural vegetation. Tourism has brought its own mistakes to the land and people. The lake still comes and goes at the whims of the rain gods. There are still tidal rhythms in people, too, ebbing and flowing into the former wild spaces along big new roads. The "juggernaut of progress" churns across the land, crushing much underneath. But the drivers prosper and there is room on top for many who can afford the ride.

We are not entitled to judge the changes in Mangola. Our hearts have layers of fondness, remorse, regret, and hope for the future survival of the indigenous people, wildlife, and land. Go visit, have some adventures yourself. May the purple onions flavor your visit.

ACKNOWLEDGMENTS

We are grateful to the United Republic of Tanzania for the privilege of living and working there.

We acknowledge the friends, lodge managers, researchers, medical people, and visitors listed below. You gave us vital logistical support, or health care, or friendship, or a context for understanding our experiences in Mangola:

Mariamu Anyawire, Dave Belden, Cheryl Bishop, Nicholas Blurton-Jones, Monique Borgerhoff-Mulder, Henry Bunn, Mary Ann Carman, Adam Chorah, Anthony Collins, Liz Cooper, Aadje Geertsema, Margaret Gibb-Kullander, Kristen Hawkes, Stephanie and Don Horal, Eddy Husslage, Glynn Isaac, Peter Jones, Saidi Kimaka, Mary and Louise Leakey, Holly Lovejoy, Audax Mabulla, Gudo Mahiya, Frank Marlowe, Fiona Marshall, Ruth Matiyas, Anne McDowell, Mike Mehlman, Joanna Mountain, Charles Musiba, James O'Connell, Pat Patton, Daudi Peterson, Mama Ramadhani, Bonny Sands, Chris, Nani and Leonie Schmeling, Daniel Sellen, Daniela Sieff, Jon and Annette Simonson, Lars Smith, Maria Strauss, Sara Tishkoff, Morimichi Tomikawa, Annie Vincent, Yoichi Wazaki, Wolfgang Weber, Richard Wrangham, and James Woodburn.

Many others in our families and scattered global network supported us. Our gratitude to you all. Special thanks to editor Julie Johnson for assistance on many levels.

FURTHER READING

Note: We met many researchers who came to study the peoples or landscape of the Lake Eyasi Basin. We encourage readers to check these people's work and websites. Their names are, alphabetically:
Nicholas Blurton-Jones, Monique Borgerhoff-Mulder, Henry Bunn, Kristen Hawkes, Glynn Isaac, Peter Jones, Mary Leakey, Audax Mabulla, Frank Marlowe, Fiona Marshall, Mike Mehlman, Joanna Mountain, Charles Musiba, James O'Connell, Bonny Sands, Daniel Sellen, Daniela Sieff, Lars Smith, Sara Tishkoff, Morimichi Tomikawa, Annie Vincent, Yoichi Wazaki, and James Woodburn.
In addition to researchers, Mangola attracted many aid workers, missionaries, journalists, filmmakers, conservation groups, tourists, and students. An extraordinary number of films, studies, articles, and books are available. See the recommendations at the end of this list to find more information.
Below is a sample of scientific reports and major books about the peoples and areas described in this book.

Blurton-Jones, Nicholas. 2016 *Demography and Evolutionary Ecology of Hadza Hunter-Gatherers (Cambridge Studies in Biological and Evolutionary Anthropology)* Cambridge University Press, UK.
Blurton Jones interweaves data from ecology, demography and evolutionary ecology to present a comprehensive analysis of the Hadza tribe.

Douglas-Hamilton, Ian & Oria. 1975 *Among the Elephants.* White Lotus Press, New York.
A very enjoyable book about what it's like to study elephants in Lake Manyara National Park; written by a husband and wife team.

Fosbrooke, H. 1972 *Ngorongoro: The Eighth Wonder.* Andre Deutsch, London.
This is the classic book on the crater, by its first Conservator.

Grzimek, Bernhard & Michael. 1960 *Serengeti Shall Not Die.* Dutton, New York.
Written by pioneer researchers and conservationists, this book has much of historical interest about the Crater and Serengeti.

Hanby, Jeannette & Bygott, David. 1981 *Lions Share.* Houghton Mifflin, Boston.
A true story of a pride of lions that share the Serengeti landscape with other creatures and with one another. A primer in ecosystem dynamics as seen through the eyes of lions.

Herlocker, Dennis. 2009 *Buffaloes by my Bedroom*. iUniverse Inc. New York.
This charming book of stories about his time in the Peace Corps in the 1960s gives one a personal feel for what it's like to work in the Ngorongoro Conservation Area.

Leakey, Mary. 1979 *Olduvai Gorge*. HarperCollins, London.
A readable basic account and good reference to the findings of the archaeological work at Olduvai carried out by Louis and Mary Leakey and their colleagues.

Leakey, Mary. 1984 *Disclosing the Past*. Weidenfeld & Nicolson, UK
Mary tells about her life as an archaeologist in this autobiography.

Matthiessen, P. 1983 *The Tree Where Man Was Born*. E.P. Dutton. Boston.
A compelling book about places and people in the Serengeti and Ngorongoro area, including a visit to the Eyasi basin. If you can, get the 1972 hardback edition with Elliot Porter's evocative photos.

Marlowe, Frank. 2010 *The Hadza Hunter-Gatherers of Tanzania*. University of California Press.
A thorough summary of findings from years of studies on the Hadza people. (The cover shows Adam, whom we meet in chapter 2 of our book)

Mlola, Gervase Tatah. 2010 *The Ways of the Tribe: A Cultural Journey Across North-Eastern Tanzania*. E & D Vision Publishing, Tanzania.
Amazingly concise and insightful overviews with illustrations of many of the tribes mentioned in our book.

Musiba, Charles, Russell H. Tuttle, Benedikt Hallgrimson, and David Webb. 1979 Swift and Sure-footed on the Savanna: A Study of Hadza Gaits and Feet in Northern Tanzania. *American Journal of Human Biology* 9(3):303 – 321
This is the report on the Footprint Man's work we watched when he came to Mangola with his colleagues.

Packer, Craig. 1994 *Into Africa*. University of Chicago Press. A personal look at primate and lion studies from the scientist, with wife Anne Pusey, who took over the Serengeti lion project from us in the late 1970s.

Peterson, Daudi. 2013 *By the Light of a Million Fires*. Mkuki na Nyota Press, Tanzania. Wonderful photos capture the liveliness while the text explores the problems and lifestyle of the Hadzabe. D. Peterson has done much to help Hadza obtain land rights and make their plight known.

Root, Alan. 2012 *Ivory, Apes and Peacocks: Animals, Adventure and Discovery in the Wild Places of Africa.* Vintage Books, Knopf Doubleday, UK.
Africa's most legendary and eccentric wildlife filmmaker reminisces about his amazing adventures with animals and people.

Sieff, Daniela F. 1997 Herding strategies of the Datoga pastoralists of Tanzania. *Journal Human Ecology* Volume 25 Issue 4 Pages 519-54.
Some results from Daniela's studies on Datoga families in the Lake Eyasi area.

Sinclair, Anthony R. 2012 *Serengeti Story: Life and science in the world's greatest wildlife region.* Oxford University Press, England.
An excellent summary of the research on the Serengeti by a scientist who has devoted at least 40 years to studying the dynamics of this wondrous ecosystem.

Stephenson, James. 2001 *The Language of the Land: Living among a stone-age People in Africa,* St. Martin's Griffin, Macmillan, USA.
A young man's account of a year spent among the Hadzabe – evocative personal descriptions of the culture and the land.

Van Lawick Goodall, Hugo and Jane. 1971 *Innocent Killers.* HarperCollins, London.
The lives of specific groups of hyenas, jackals, and wild dogs living in Ngorongoro Crater and on the plains are described and illustrated. Fascinating.

PHOTOGRAPHIC BOOKS
Beckwith, C. and Ole Saitoti, T. 1980 *Maasai.* Abrams, New York.
A photographic study of the Maasai with text written by a Maasai born and raised in the Ngorongoro area. He has also written an autobiography which deals more specifically with the problems faced by a Maasai in coping with Western culture.

Ole Saitoti, T. 1986 *Worlds of a Maasai Warrior.* Random House, New York.
We knew the author; his life was very much that of insider-outsider, a man of two cultures.

Iwago, M. 1987 *Serengeti.* Thomas and Hudson, London; 1984 Asahi Shimbun, Tokyo.
One of the most beautiful photographic studies of Serengeti and Ngorongoro wildlife.

Johns, Chris. 1992 *Valley of Life.* A lovely book of photos taken by the National Geographic photographer in East Africa's Great Rift Valley.

Künkel, Reinhard. 1992 *Ngorongoro.* Harvill, London.
Some amazing photos, from massive "sculptural" studies of rhinos, elephants and buffalos to intimate views of smaller creatures. The text is full of interest and humor.

Lithgow, Tom and Van Lawick, Hugo. 2005 *The Ngorongoro Story.* Camerapix, Kenya
Stories of the life of settlers and hunters in the Ngorongoro area. Photos by Hugo van Lawick, drawings by David Bygott.

Van Lawick, Hugo. 1977 *Savage Paradise.* HarperCollins, London and 1986 *Predators and Prey*, Random House, New York.
Two pictorial books with dramatic scenes and unusual events in the lives of predators and their prey; photographed mostly in Ngorongoro and Serengeti. Van Lawick also wrote *Solo, the story of an African wild dog*, with illustrations by David Bygott.

Woodburn, James. 1970 *Hunters and gatherers: the material culture of the nomadic Hadza.* British Museum, UK
The book, if you can get hold of it, describes Woodburn's work from 1958-1969. It has pictures that show actions, items, and species that the Hadza relate to in their daily lives. Woodburn still visits the Hadza and is a repository of tales and insights.

WEBSITES
Use a search engine to find information about the major ethnic groups of the Eyasi area: Hadza, Iraqw, Datoga, Maasai and Bantu tribes (e.g., Sukuma, Isanzu, Nyaturu).
Wikipedia is a good source for information about tribes and places mentioned in this book.
Check the official websites for Ngorongoro Conservation Area, Serengeti, Manyara, and Tarangire National Parks for up-to-date information.
If you are interested in safaris, there are innumerable websites to entice you. But check out the companies on sites such as TripAdvisor before you commit, as some promise more than they can deliver.

ABOUT THE AUTHORS

Jeannette Hanby has taken on jobs ranging from working with abused and abandoned children in Los Angeles County to placer gold mining in the Sierra Nevada mountains. After completing her Ph.D. dissertation in Oregon on monkeys, she went to Cambridge in England for further study of primates—one of whom, David Bygott, became her husband.

David Bygott began observing wildlife and drawing as a child exploring the countryside of southern England. He was finishing his Ph.D. thesis on wild chimpanzee behavior when he met Jeannette, his partner in field research, conservation, and business for almost five decades.

The couple studied lion biology in Tanzania's renowned Serengeti National Park and Ngorongoro Conservation Area. They have worked as conservation educators, university lecturers, safari guides, writers, and artists. Together, Jeannette and David have produced numerous guidebooks, educational booklets, activity books, and museum displays in both English and Swahili.

They now live in an intentional community in Tucson, Arizona, still learning about primate behavior, as well as continuing to travel and explore the world.

Made in the USA
Las Vegas, NV
13 November 2021